D1405094

# Toward a Solar Civilization

The MIT Press
Cambridge, Massachusetts, and London, England

**Toward a Solar Civilization**
**Edited by Robert H. Williams**

Acknowledgments
The papers in this volume originally appeared in the *Bulletin of Atomic Scientists:* chapter 1, November 1975; chapter 2, February 1976; chapter 3, March 1976; chapter 4 (originally titled "Public Policy for Solar Heating and Cooling"), October 1976; chapter 5, May 1977; chapter 6, January 1976; chapter 7, April 1976; chapter 8, September 1976; chapter 9, May 1976; chapter 10, May, 1976; chapter 11, June, 1976; chapter 12, November 1976; and chapter 13, October 1977.

This book was set in IBM Theme by To the Lighthouse Press. It was printed on R & E Book and bound by Murray Printing Company in the United States of America.

Library of Congress Cataloging in Publication Data

Main entry under title:

Toward a solar civilization.

Includes index.
1. Solar energy. I. Williams, Robert H., 1940–
TJ810.T68        333.7        78-16567
ISBN 0-262-23089-5

Preface    vii

I    Introduction

1    Solar Technologies    1
Frank von Hippel and Robert H. Williams

II    Solar Space Heating

2    Active-Type Solar Heating Systems for Houses: A Technology in Ferment    17
William A. Shurcliff

3    Why Not Just Build the House Right in the First Place?    33
Raymond W. Bliss

4    Barriers and Incentives to Encourage the Use of Solar Heating and Cooling    51
Alan S. Hirshberg

III    Solar Electricity

5    Solar Power Plants: Dark Horse in the Energy Stable    73
Richard S. Caputo

6    Solar Sea Power    95
Clarence Zener

7    Photovoltaic Solar Energy Conversion    107
Martin Wolf

8    Wind Energy    125
Bent Sørensen

IV    Fuels from Solar Energy

9    Flower Power: Prospects for Photosynthetic Energy    145
Alan D. Poole and Robert H. Williams

10    Tower Power: Producing Fuels from Solar Energy    169
Michael J. Antal, Jr.

# Contents

# V Development Strategies

11 Solar Energy and Rural Development for the Third World    181
Arjun Makhijani

12 Energy Efficiency: Our Most Underrated Energy Resource    203
Marc H. Ross and Robert H. Williams

13 Toward a Solar Civilization    221
Frank von Hippel and Robert H. Williams

Contributors    241

Index    243

Solar energy should not be approached with cool indifference, because solar energy is not simply just another energy source. Its properties are such that we should want very much to make solar energy a major energy resource. In one form or another solar energy is available everywhere; it is renewable; and in many cases (but not all—as the first paper in this book makes clear) its exploitation is kinder to the environment than is the case with conventional energy forms. In addition, the prospects for organizing much solar technology on a human scale is attractive to those who are uncomfortable with the long-established trends toward large-scale energy technologies and the associated centralized bureaucracies.

But is solar energy practical, economical? This question does require some dispassionate analysis. Despite their attractions, most solar technologies are not economically competitive with conventional energy technologies today because solar energy is diffuse and intermittent, requiring rather costly collection and storage systems. The attractions of a solar world have made some solar enthusiasts insensitive to the practical problems that must be overcome in order to make solar energy affordable. At the same time, however, those analysts who concern themselves mainly with conventional fossil and nuclear energy sources often tend to dismiss solar energy with a cavalier and, I believe, foolish disregard of the potential for innovation in solar technology.

Interest in solar energy has been spurred by the rapid rise in energy prices; by the prospect of running short of oil and gas resources; and by the serious environmental, safety, and security problems that have clouded future prospects for coal and nuclear energy sources. In part this interest in solar energy is reflected in a rapidly expanding federal research and development budget for solar energy. In 1972, the National Science Foundation was the lead agency for solar energy research in the United States, with a budget of $2 million per year. By fiscal 1977, the federal solar research and development budget had grown to $290 million. At least as important as this federal effort is the burgeoning growth of a community of self-selected solar technologists and entrepreneurs. It is striking that many of the more innovative and promising ideas have emerged from small groups with restricted budgets operating outside the energy research "establishment."

The papers in this collection were prepared by specialists and policy analysts for a solar energy series of the *Bulletin of the Atomic Scientists* (1975–1977). One important message conveyed by these papers is that there would be a rich diversity of technologies in a solar-energy- powered

# Preface

economy. There would be many ways to perform a given task, and there would be marked regional variations in the solar energy supply.

Consider for example the many different strategies for heating a house. Most of the solar heating effort to date has been devoted to the design and development of rooftop solar collectors and the associated heat storage units. As Shurcliff points out, there is a large and growing number of such systems of varying complexity and cost. Such systems do not exhaust the possibilities, however. Bliss describes a promising alternative: a directly (passively) heated solar house where the solar collector and storage units are integral parts of the house structure. Still another approach would be to recover, for space heating, waste heat from power generation in community-level solar energy systems, as suggested in the last chapter in this volume.

While today's commercial energy systems tend to be similar the world over, there would be marked regional variations in the mix of deployed solar technologies to reflect regional differences in the solar energy resource. The thermal-electric and photovoltaic systems described by Caputo and Wolf would best be deployed in dry, sunny areas like the Southwest; the production of chemical fuels from agricultural waste (see Poole and Williams) in the corn belt of the Midwest; windmills (see Sørensen) in the Great Plains, where there are strong, steady winds; and ocean thermal gradient generators (see Zener) in tropical oceans.

A principal source of excitement for solar technologists is the fact that the potential for innovation in solar technology is very great. Shurcliff captures the sense of excitement accompanying the intensely creative ferment that is sweeping the solar-heating field. Similarly, in the biomass area many promising technologies are being proposed for making useful chemical fuels from biomass feed stocks: biogasification and cogeneration (Poole and Williams) are examples of technologies for utilizing organic waste that could be developed in the near term; the "tower power" concept proposed by Antal is a promising intermediate-term technology for stretching the fuel value of limited organic waste resources.

Broad policy issues will be at least as important as technological considerations for the successful development of solar energy. Hirshberg describes some of the institutional obstacles to widespread implementation of solar-heating and -cooling technology and suggests policies for overcoming them. Makhijani argues that appropriate solar technologies for rural villages in the Third World are those that would be compatible with the resources available in those areas and supportive of economic development there.

One of the least-appreciated aspects of a solar energy strategy is that it must be accompanied by a strong emphasis on eliminating energy waste. As the Bliss paper shows so well for the directly heated solar house, more efficient energy use and the exploitation of solar energy must be complementary considerations in an integrated design strategy for energy supply and demand. To show what the prospects are for more effective energy utilization, the paper by Ross and Williams on the potential for energy conservation has been included in this volume.

The reader will not find in these essays a blueprint for a transition to a solar civilization. Indeed, the authors represented here often hold conflicting views as to the prospects for particular solar technologies—as one would expect in an embryonic and fast-moving field. But I hope that the reader will be infected by the excitement of the challenges facing solar energy technologists and will also be stimulated to think about the implications of what success in these endeavors would mean for the future of society. Many of the more promising strategies for recovering solar energy involve arrangements for energy supply and demand that are largely unfamiliar in today's energy systems. As the final essay speculates, "a shift to a solar economy might reshape our way of life at least as profoundly as did the introduction of the automobile or the products of the electronics industry."

# I Introduction

In the face of the increasingly pessimistic outlook for oil and gas availability, environmental concerns about coal-fired and nuclear power plants and rising energy prices, solar energy is being looked upon as a potential major energy source for the long-term future. The purpose of this article is to briefly survey some of the basic characteristics of alternative solar conversion systems so as to help show the inherent relative promise and limitations of different systems.

Solar energy (in the form of hydroelectric power) today supplies only about 4 percent of the 75 quadrillion Btu of energy consumed in the United States.[1] The Energy Research and Development Administration sees the solar share expanding considerably in the decades ahead, until perhaps 25 percent of the nation's energy would come from solar technologies by the year 2020.[2] However, it is interesting to note that our reliance on solar energy is much greater than these figures suggest. If the definition of our energy budget is broadened to include all the energy sources that power our life support system, then solar energy is already our dominant energy source.

Consider first photosynthesis, the process by which the earth's inhabitants are provided with food, fiber, wood and pulp products as well as the forests, meadows, and cultivated greenery that we enjoy in outdoor recreation. While the solar energy converted to plant matter in continental United States each year amounts to about 100 quadrillion Btu, the amount of sunlight used in this process is actually much greater. Plants must transpire some 500 pounds of water for each pound of dry plant matter produced. The solar energy input for this evapotranspiration is on the order of 10,000 quadrillion Btu or 20 percent of the solar energy falling on the United States each year.

An even greater amount of solar energy, about 13,000 quadrillion Btu, is used in evaporating (and in the process desalinating) seawater to provide most of the 30 inches average annual rainfall in the United States. This rainfall gives us fresh water for direct human and animal consumption; it gives us fresh water for our rivers, lakes and streams for transportation and outdoor recreation; and it provides for the transpiration needs of plants.

Overall, evaporation of water from oceans and from the land requires on the order of 20,000 quadrillion Btu of solar energy per year for the United States. All other energy inputs are miniscule in comparison. Thus, the "official" U.S. energy budget, dominated at present by fossil fuel, amounts to only about 0.4 percent of the total energy supply if the total energy budget is redefined to include these solar inputs.

---

# 1 Solar Technologies
## Frank von Hippel and Robert H. Williams

Besides showing the scale of our present dependence on solar energy, these numbers help convey a sense of the immensity of the solar energy resource. But while the solar energy arriving at the surface of the 48 contiguous states is about 600 times the level of commercial energy consumption, it is also both diffuse and intermittent. The average rate at which solar power is received in the United States per unit area of horizontal surface (the average insolation) is only about 17 watts per square foot.[3] (Insolation includes both direct ("incident") and diffuse solar radiation. The solar insolation at the earth's surface varies with latitude, season, cloudiness, and the dust in the atmosphere as well as the time of day.) Gathering sunlight and providing it in useful form when needed at competitive cost is the principal challenge to solar energy technology.[4]

### Residential Use

There is a growing consensus among both solar energy advocates and energy policy analysts that solar energy will first be widely used in the United States to provide water and space heating in homes and in some commercial applications. One of the principal reasons for this optimism is the comparability of the solar energy supply at the household level with the residential demand for heat. In the winter, for example, a typical U.S. home directly consumes fossil fuel for space and water heating at an average rate of about 8 kilowatts. For comparison, the average winter solar power incident on a 1,000-square foot south facing roof top collector would be about 14 kilowatts. Thus a 30 to 40 percent efficient collector would be able to provide 50 to 70 percent of this heating demand. If fuel conservation efforts were pursued simultaneously, the area of the collector required could be reduced.

This comparability of supply and demand extends to electricity, where there is the possibility that photo-electric cells may one day be widely used to convert sunlight directly to electricity in rooftop units. (Photoelectric cells or solar cells are solid-state devices which collect the radiant energy of the sun and convert it directly into electric power.) A typical household today consumes electricity at an average rate on the order of 1 kilowatt. This could be supplied by a 10 percent efficient photoelectric unit covering 600 square feet of a south facing roof.

The principal obstacle to the immediate widespread adoption of solar energy for homes is the high capital cost—mainly of the collector and storage system. In order to provide solar energy when the sun doesn't shine, a

Von Hippel and Williams

storage system is required—for example, hot water or hot rocks for solar heat and batteries for solar electricity. Capital costs limit the period of overcast skies for which it is economically advantageous to provide storage. To store enough heat for a winter week of overcast skies would require a very well insulated water tank with about 1,000 cubic feet of capacity for the household that consumes fossil fuel at an average rate of about 8 kilowatts. And to store a week's supply of electricity would require several tons of lead-acid storage batteries for the household that consumes electricity at a rate of about 1 kilowatt.

The large extra costs involved in designing solar storage systems to meet infrequent peak demands have led designers to focus on schemes with thermal storage capacities of perhaps 2 days and electrical storage capacities of less than a day. Back-up fuel burning heating systems and a connection to the electrical utility are therefore required.

The capital cost issue for solar energy can be expressed in terms of the economically allowable cost per square foot of collector. Over a 6 month heating season a 40 percent efficient collector could save, for each square foot, the equivalent of three-fourths of a gallon of heating oil (30 cents) or 20 kilowatt hours (80 cents) in the case of a house with electric resistance heat. To realize these fuel savings, however, a prospective homeowner may have to pay more for a solar home. A total additional investment, including storage, of $4 per square foot as a replacement for oil heat or $12 per square foot as a replacement for electric resistance heat would be justified at present fuel prices for a 20-year mortgage at 9 percent interest.[5] The prospects for reducing capital costs to $4 per square foot for systems using rooftop water cooled solar collectors are not good for most of the designs currently being promoted. However, simplified designs such as those used in houses built by H. Thomason in the Washington, D.C., area may already be in this cost range.[6]

One very interesting possibility for low cost space heating is a house with solar components serving other useful functions. Solar energy, captured by south facing windows, provides over half of the space heating requirements for the house shown in figure 1.1.[6] Massive walls and floors provide the heat storage system, and very careful insulation reduces losses. Slightly recessing the south facing window will exclude the direct sun in the summer but not in the winter. Houses such as the one shown in figure 1.1 underscore the importance of integrating solar collection and storage into the building design from the beginning.

Before photoelectric cells can produce electricity competitively, con-

**Figure 1.1**
Solar home. This house in Weston, Mass., was built in 1960. With enclosure of the overhang in front of the basement, about 60 percent of the heating in the winter of 1974–1975 was provided by solar energy. Warm air is distributed throughout the house by electric fans. (See n. 6.)

siderable capital cost reductions will have to occur. If electricity costs twice as much as it does today (not an unlikely development over the next couple of decades), an investment of $20 per square foot or about $2 per (peak) watt would be justified by the homeowner. At present the cost of solar cells alone (exclusive of energy storage, support system, and installation costs) is about $10 per (peak) watt. With concentrators this cost gap can be considerably reduced. Moreover, progress is rapid in this area, and substantial cost reductions through technological innovation and mass production techniques are expected over the next decade or so.

## Thermal Electric Plants

Fuel-fired electrical generating plants are based on the conversion of the chemical or nuclear energy of the fuel into heat energy which is then used

Von Hippel and Williams

to run a heat engine, ordinarily a steam driven turbogenerator. Many concepts for generating electrical power from solar energy are based on this same basic technology.

*Land-Based* This concept depends upon the solar energy being absorbed and converted into heat using large land-based collectors. It would be attractive to locate such systems in areas where there is a great deal of available space and year-round sunshine. The southwestern United States is a likely area, with average insolation in December about twice the national average.

Unfortunately, like other power plants, solar-thermal electric plants using a steam cycle would require large quantities of cooling water, and water is in short supply in the Southwest. Therefore, designs which concentrate the solar heat at high temperatures (with the associated high efficiencies for converting heat energy into electrical energy), and designs involving rejection of waste heat directly to the air would be favored.

A 1,000-megawatt (electrical) plant in the Southwest would require a collecting area of about 15 square miles if it could convert the incident solar energy to electricity at an overall efficiency of 10 percent. The only comparable commitment of land for a generating station of this size is for a hydroelectric plant with its enormous reservoir. (It should be noted, however, that over the 30-year life of a coal-fired plant with equal output an equal amount of land would have to be surface mined for the coal if the coal seam averaged 7 feet in thickness.)

Some of the principal economic challenges to land-based thermal-electric energy systems are the same as those facing residential heating systems: bringing down the costs of collectors and energy storage.

*Ocean-Based* This concept uses the tropical or subtropical ocean surface as both a free energy collector and a heat storage device. The basic idea is to make use of the temperature difference (typically $25^\circ$ F or more) between warm surface waters and cold bottom water in these regions to run a heat engine for electrical energy production. Because of the low temperature of the heat source, the ocean thermal plant is very inefficient, converting only about 2 percent of the heat extracted from surface waters to electricity.

Heat exchanger design is one of the principal technological challenges in this concept. To operate the heat engine most effectively, within the constraints of the small temperature differences involved, requires exchanging heat between the ocean waters and the working fluid of the heat engine with the least possible change in temperature. Heat exchangers with large surface areas are required to accomplish this. These heat exchangers dominate the capital cost of this system. In research and development particular attention is being given to developing special surfaces with improved heat transfer properties.

The Gulf Stream is a potentially attractive site for establishing an ocean thermal gradient system. This is because the useful energy flowing through a cross-sectional area is concentrated by about a factor of 1,000 relative to that of the average solar insolation on land, even with the extracted heat limited to what is available with a 1° F temperature drop in the warm surface water passing over the heat exchanger.

If we extracted too much energy from the Gulf Stream, we would seriously affect the climate of Western Europe. This would probably not happen, however, until exploitation reached quite a high level. The current electrical energy demand of the United States could be met from this source with a temperature drop of the Gulf Stream of only about 0.5° F. It is not known whether this would be "too much."

An alternative which would not raise this concern would be to build plants which "graze" relatively still areas such as in the Gulf of Mexico. A 1,000-megawatt (electrical) plant, or group of plants with this capacity, would have to graze about a 1,000 square mile area in order to sustain a surface temperature drop no greater than 2° F. This area is so large because more than 90 percent of the incident solar energy would be lost by the ocean directly to infrared radiation and evaporation—the ordinary cooling processes of the ocean.

Some proponents of the ocean thermal gradient scheme suggest that a major side benefit could accrue from the nutrients brought up from the ocean depths along with the cooling water. Natural upswellings of deep ocean water (off Peru for example) are associated with rich fisheries. It should be pointed out, however, that bringing this cold water to the surface in this variant on the ocean thermal gradient scheme could also cause a net transfer to surface waters of inorganic carbon from the carbon-enriched bottom waters.

Most of the excess carbon would pass into the atmosphere as carbon dioxide.[7] In the absence of compensating effects, if the cold water is not returned to the depths, an ocean thermal gradient plant would release

Von Hippel and Williams

about one-third as much carbon dioxide to the atmosphere per kilowatt hour of electricity produced as a fossil-fueled plant. Since the buildup of carbon dioxide in the atmosphere from the burning of fossil fuels is a major environmental concern,[8] it would be unfortunate if solar energy were developed in a manner which would exacerbate this problem.

A potentially serious constraint on this concept may arise if ocean thermal gradient plants are developed on a large scale. Except for a few sites near islands, such as Hawaii and Puerto Rico or off the coast of the Southeast United States, the difficulties of long distance underwater transmission of the generated electricity would tend to discourage utility interest. Proponents have suggested that if the power became sufficiently cheap, electricity could be converted into more transportable energy forms such as hydrogen. Alternatively the transmission problem could be avoided if energy intensive industries were located near ocean thermal gradient sites, just as, for example, aluminum refineries have been remotely sited in the past to take advantage of low cost hydroelectric power.

Similar energy transmission problems would also arise for a large land-based solar development in the Southwest or a wind-electric system in the Aleutians. Conversion of cheap solar energy into an easily transportable form therefore deserves study in its own right.

It should be emphasized, however, that there are sites where the energy transmission problem would not be serious and that initial economic studies of the ocean thermal gradient concept are encouraging. Further development of this technology with a pilot plant at a suitable location therefore appears to be warranted.

## Wind and Wave Power

The step involving the greatest energy losses in the conversion of solar energy to electrical energy in thermal systems is that step in which the heat is converted to mechanical energy. It is of considerable interest therefore that thermal energy originally from the sun is made available as mechanical energy by natural processes on a large scale, notably in the form of wind and waves. About 1 percent of the solar influx generates the great atmospheric pressure systems which drive the winds which, in turn, generate the waves.

*Wind Power* The energy associated with wind can be more concentrated than the original solar influx from which it arises. At a favorable location,

**Figure 1.2**
Two approaches to wind power. Left: 10,000 water pumping windmills cover a plain on the island of Crete. Right: a huge (6.5 megawatt electrical) wind power generator dwarfs a farmhouse in this sketch illustrating a proposal made in 1945 by Percy Thomas of the Federal Power Commission. (See "Solar and Aeolian Energy," *Proceedings of International Seminar on Solar and Aeolian Energy*, Sounion, Greece, Sept. 4–5, 1961; and n. 9.)

a typical wind velocity at a couple hundred feet above ground level is 20 miles per hour. At this velocity the power through a surface perpendicular to the wind is 45 watts per square foot, or nearly 3 times the average solar insolation on the ground. The high quality of this mechanical energy is manifest in the fact that a windmill can convert to electricity about one-third (ideally, 16/27 or about 59 percent) of the energy incident on the area swept out by the propeller. Thus a wind of 20 miles per hour would yield nine times as much electric power per unit area as a 10 percent efficient solar thermal unit.

An important characteristic of wind is that the power developed increases as the cube of the wind velocity, so that the power available at 25 miles per hour is about double that available at 20 miles per hour. This

Von Hippel and Williams

means that the average power developed by a windmill can be considerably greater than the power produced from a wind of average speed.

Large-scale development of wind power would lead to a proliferation of windmills. To get a feeling for the scale involved, consider a system of 200-foot diameter windmills. Replacing all of today's electrical generating capacity in the United States with windmills would require arrays of some one million of such windmills which, if lined up side by side, would stretch 40,000 miles. The number of windmills involved is on the same order as the number of transmission towers for high tension lines in the United States. It is not obvious that the climatic impact of such intensive use of the wind would be negligible. Aesthetic considerations would also have to be taken seriously, as a comparison of the windmills in figure I.2 suggests.

Windmill power would appear to require energy storage for times when the wind is not blowing. However, as Percy Thomas [9] pointed out in his classic 1945 study on wind power, when the wind dies down in Philadelphia it may be blowing in New York. The required storage capacity can be reduced considerably therefore by feeding electricity from dispersed windmills into a regional electric power transmission grid. Thomas did some

Solar Technologies

preliminary surveys which suggested that such an arrangement would substantially reduce the fluctuations in wind-generated power. A high priority for wind power research should be to pursue such studies in greater depth for specific proposed distributions of wind generators.

The engineering knowledge required for optimized windmill design is mature relative to the state of development for other types of solar electric power plants. It should be possible to decide in the near future whether or not deployment of windmills is economically justified. Preliminary capital cost estimates seem to be "in the right ball park" to justify the deployment of full scale demonstration plants at appropriate sites.[10] Moreover, the relatively small unit sizes of windmills and the corresponding large numbers needed to produce appreciable generating capacity make windmills well suited for cost cutting mass production techniques.

*Wave Power* For a wind of about 20 miles per hour, a fully developed "sea" in the open ocean has waves about 5 feet high from trough to crest. Such a sea carries an average power of about 5,000 watts per linear foot,[11] which is comparable to the wind power intercepted (per foot of propeller diameter) by a large windmill operating in similar wind conditions. Wave power, like wind power, is high quality mechanical power. The potential for harnessing wave power should be assessed and the economics compared to wind power.

## Solar Satellites

While there do not appear to be substantial scale economies for earth-based photoelectric converters, one glamorous proposal for harnessing the sun involves converting sunlight to electricity with photocells at a space station in geosynchronous orbit.[12]

Putting these power stations in orbit would require great effort, however. If the United States (or the world) should take this option seriously, it is reasonable to expect that these plants would come on-line at least as fast as some 10,000 megawatts (electrical) per year. (This corresponds to 2½ percent of U.S. generating capacity today.) Putting 10,000-megawatts (electrical) of space station generating capacity into operation per year would involve at least two space shuttle flights per day, given a payload capacity of 65,000 pounds and an estimated space station weight of 5 pounds per kilowatt.[12]

Such intensive shuttle activity could have harsh environmental impacts. Chlorine from rocket exhaust is especially worrisome, given the intense concern lately about the effects of chlorine on the stratospheric ozone layer.[13,14] A recent environmental impact study of the space shuttle[15] showed that over 100 tons of chlorine would be dumped directly into the stratosphere as hydrochloric acid for each shuttle flight. At two shuttle flights per day, chlorine would be introduced throughout the stratosphere as fast as the maximum rate of stratospheric chlorine production in 1980 likely to result from the release of freon at ground level.[13]

This would probably be unacceptable. Other technological options exist for putting these stations in orbit but they are not being developed. This calculation points up a potential serious impact of space shuttle operations generally on the ozone layer.[15] It also demonstrates the importance of closely scrutinizing all parts of a system in a feasibility study.

## Photosynthesis

Photosynthesis is attractive as a solar energy conversion process because, unlike most solar schemes, it provides in plant material a natural, convenient energy storage medium. Moreover, burning plant material, unlike fossil fuels, would not lead to a net atmospheric build-up of carbon dioxide and the climatic consequences implied by the build-up because, on the average, plants would be grown as fast as they are burned. But photosynthesis is the least efficient solar conversion system commonly considered. Theoretically, photosynthesis has only a 5 to 6 percent maximum net efficiency for the conversion of natural sunlight into chemical energy. Practical efficiencies are far less. In tropical bogs nature is at her best with a 4 percent photosynthetic conversion efficiency. Typical annual average values for U.S. agriculture are in the range 0.25 to 0.75 percent.

One hundred years ago the conversion of solar energy to organic materials through photosynthesis provided the primary source of power and heat for the population of the United States in the form of fuel wood and work animal feed. Since that time, however, the U.S. population has increased about 4.5 fold and our per capita energy use has increased about 2.5 fold[16] for a total growth in energy consumption of more than a factor of 10. Accordingly, while photosynthesis can potentially provide a significant fraction of total energy needs, with low photosynthetic conversion efficiencies it would appear unlikely that we can "go back to the good old days" to a complete reliance on this source because of land use constraints.

Because efficiencies are so low, a sensible approach to using photosynthesis for energy production would be to use by-products of other economic operations based on photosynthesis. A prime candidate is our food production system. The incentive to grow crops for food production is clearly greater than that for growing crops for fuel: the farmer gets about $45 or more per barrel of oil equivalent energy in the wheat he sells as food, or 4 times the price of a barrel of imported oil.

However, there is a great deal of energy-rich (if not food-rich) organic material generated as a by-product of our food production system. The harvest taken from U.S. fields represents, on the average, only about one-half the mass of the plants (dry weight exclusive of roots). In addition, a significant fraction of our grains and soybeans are used as animal feed, and typically 25 to 30 percent of the food energy ends up in excrement.

A recent study by Poole[17] provides a careful analysis of the energy resource potential of these and other organic wastes. Of the 690 million tons (dry weight) of organic wastes produced each year in the United States, about 340 million tons are crop residues and 200 million tons are manure. At 16 million Btu per dry ton these agricultural wastes represent a resource base of some 8.6 quadrillion Btu per year (11 percent of U.S. energy consumption in 1973). Not all of these wastes are recoverable, however. Economic constraints limit crop residue recovery below some minimum areal density and only a fraction of the manure is available in feedlots where it is sufficiently concentrated for economic recovery.

Losses also occur in converting the energy to a convenient form. One especially attractive conversion process, the production of methane via bacterial digestion under anaerobic conditions, has an efficiency of 37 to 55 percent. For this process, plant nutrients remain concentrated in the residuum, which can either be returned to the soil as fertilizer or converted to a protein feed source for animals. Poole finds that, with the various losses taken into account, the by-product methane potential from our food system would amount to 2.5 to 4.0 quadrillion Btu per year, corresponding to 10 to 20 percent of the current natural gas production rate. Unlike natural gas wells these wells would not run dry.

In this brief overview we have attempted to show how the varied technologies for exploiting solar energy compare to one another. In such limited space our analysis of alternative options is necessarily oversimplified. The basic point we wish to make is that a good understanding of the physical advantages and limitations of different options is necessary to provide a solar research and development program with a rational basis. In particular, systems studies of the implications of large scale development of the alternative schemes would be helpful in shaping the solar energy program.

With a few examples we have tried to dispel the popular notion that so-

Von Hippel and Williams

lar energy is inherently pollution free. All energy systems involve some pollution, and solar energy is no exception. We have suggested how potential impacts could range from the atmospheric buildup of carbon dioxide to ozone destruction. Resource availability, such as water for cooling, may prove limiting for some options. With throughtful system design much can be done to minimize such environmental and resource problems. However, these considerations underscore the importance of maintaining a diversity of conversion concepts in developing solar technology, so as to avoid reaching "limits" for any particular concept. These considerations also point up the importance of coupling solar energy development to aggressive energy conservation programs. If solar energy is to capture a significant fraction of our energy budget, our energy budget must not be too large.

Compared to other energy technologies the capital costs of solar energy are often high. But solar energy is nevertheless of great interest in our society today, as we come to the end of an era of "cheap" fossil fuels, with no assurance that we are entering a new era of "cheap" nuclear energy. Moreover, solar energy is a resource with potentially great promise, especially in the forms of wind power and photosynthetic energy, for the energy hungry developing nations, where conditions for exploiting this resource are often favorable and per capita energy consumption is low.

### Notes

1. One BTU, or British thermal unit, is the amount of heat required to raise the temperature of one pound of water approximately $1°$ F; 1 quadrillion Btu = $10^{15}$ Btu, is equivalent to roughly 40 million tons of coal or 180 million barrels of oil.

U.S. Bureau of Mines, "U.S. Energy Use Down in 1974 After Two Decades of Increases," News Release (Washington, D.C.: The Bureau, April 3, 1975).

2. U.S. Energy Research and Development Administration, "National Solar Energy Research, Development, and Demonstration Program, Definition Report" (Washington, D.C.: ERDA, Aug. 6, 1975).

3. This is only about 13 percent of the solar flux in space near the earth. The space solar flux is reduced by a factor of four on the average due to the fact that the area of the earth's disk perpendicular to the sun is only one-fourth the total surface area of the earth.

A further reduction averaging approximately a factor of two over the globe results from reflection from clouds, dust and the individual molecules in the atmosphere and from absorption in the atmosphere. Aside from atmospheric conditions, the solar flux on a horizontal plane at the earth's surface depends on the latitude.

4. For an excellent introduction to a large number of 'do-it-yourself' solar projects, see: *Energy Primer: Solar, Water, Wind and Biofuels* (Menlo Park, Ca.: Whole Earth Truck Store/Portola Institute, 1974).

Solar Technologies

5. A 9 percent interest rate in current dollars corresponds to a 3.3 percent interest rate in constant dollars, for an inflation rate of 5.5 percent.

6. For an excellent descriptive listing of solar heated buildings, see W. Schurcliff, "Solar Heated Buildings: A Brief Survey," which is available from the author (19 Appleton St., Cambridge, Mass. 02138).

7. Surface waters contain about 3.5 milligrams per liter less inorganic carbon than water below 1,000 meters. A buffering action in surface waters gives rise to the phenomenon that the fractional increase in the partial pressure of carbon dioxide is much greater than the fractional increase in the total inorganic carbon concentration. Hence, most of the excess inorganic carbon brought into the surface layer passes into the atmosphere as carbon dioxide. See. R. H. Williams, "The Greenhouse Effect for Ocean Based Solar Energy Systems," Working Paper No. 21 (Princeton, N.J.: Center for Environmental Studies, 1975). For a recent detailed discussion of the exchange process for carbon dioxide between the ocean and atmosphere, see Charles D. Keeling, "The Carbon Dioxide Cycle: Reservoir Models to Depict the Exchange of Atmospheric Carbon Dioxide with the Ocean and Land Plants," in *Chemistry of the Lower Atmosphere*, ed. by S. I. Rasool (New York: Plenum Press, 1973).

8. Evidence is presented in a recent article that suggests that the present global cooling trend may be one phase of a long-term climatic cycle which can be expected to end soon. Subsequently we should expect a pronounced global warming from increased carbon dioxide in the atmosphere. For details, see Wallace S. Broecker, "Climate Change: Are We on the Brink of a Pronounced Global Warming?" *Science*, 189: 4201 (August 8, 1975), 460–463.

9. Percy H. Thomas, *Electric Power from the Wind* (Washington, D.C.: Federal Power Commission, 1945).

10. See, for example, Bent Sørensen, "Energy and Resources," *Science*, 189:4199 (July 25, 1975), 255.

11. See Willard Bascom, *Waves and Beaches* (Garden City, N.Y.: Doubleday, 1964), pp. 53, 244. Wave power increases slightly more rapidly than the wave height squared.

12. Peter E. Glaser, "Space Solar Power: An Option for Power Generation," paper presented at 100th Annual Meeting of the American Public Health Association, Atlantic City, N.J., Nov. 14, 1972.

13. R. J. Cicerone, et al., "Stratospheric Ozone Destruction by Man-Made Chlorofluoromethanes," *Science*, 185:4157 (Sept. 27, 1974), 1165–1167.

14. S. C. Wofsky, et al., "Freon Consumption: Implications for Atmospheric Ozone," *Science*, 187:4176 (Feb. 14, 1975), 535–537.

15. R. J. Cicerone, et al., "Possible Environmental Effects of Space Shuttle Operations," interim report to the National Aeronautics and Space Administration, prepared at the Space Physics Research Laboratory, University of Michigan, Ann Arbor, May 1973.

16. John C. Fisher, *Energy Crisis in Perspective* (New York: Wiley, 1974).

17. Alan Poole, "The Potential for Energy Recovery from Organic Wastes," in *Energy Conservation Papers*, ed. by Robert H. Williams (Cambridge, Mass.: Ballinger, 1975).

# II    Solar Space Heating

Most of the efforts being made today to promote solar heating of houses are focused on the so-called active type of solar heating system, that is, systems employing a flow of water or air to carry the heat from the solar-radiation collector to the storage system or directly to the rooms of the house. Much less effort is devoted to *passive*-type solar heating, where the sun's rays impinge directly on floors, walls, etc., and no channeled flow of water or air is needed.

At the end of 1975 active-type solar heated houses outnumbered the passive-type houses almost 10 to 1: there were about 140 of the former and about 16 of the latter.[1] A solar-heating industry is developing rapidly. Hardly a week goes by without the formation of a new company eager to sell collectors or associated equipment. Several giant corporations, too, have tossed their hats into the ring. Already there are over 40 companies in the United States offering water-type collectors for the solar heating of buildings, and there are a dozen offering air-type collectors.[2] About 100 companies are selling small water-type collectors for supplying heat to domestic hot water systems. Many new designs of collectors are being tried out. The streams of announcements, reports, brochures, operation manuals, etc., has become a mighty river.

A few schemes seem to me to be genuinely successful in terms of performance and cost. Most of these are off-beat schemes, however, which appear crude in some respects and many architects and builders have shied away from them. Most types of collectors are too expensive or are of questionable durability.

Even though the industry is growing, much gloom is evident. There are many companies eager to produce collectors, but most such companies make few sales. Each company is struggling to modify its designs so as to achieve better performance, greater durability, lower cost. Many hurdles have yet to be surmounted.

The principle of operation of a typical active-type solar-heating system is well known. The sun's rays pass through two sheets of glass, strike a black-coated sheet of metal, and are absorbed there. The heat in the metal is picked up by a stream of water (the coolant) and carried to an insulated water-filled tank in the basement (or is picked up by a stream of air and carried to a huge bin filled with stones). When the rooms of the house become cold, heat is drawn from the tank or bin and delivered to the rooms by conventional means: radiators, forced hot air, etc. Detailed descriptions and analyses of such systems have been presented in hundreds of articles and a few books.[3]

# 2  Active-Type Solar Heating Systems for Houses: A Technology in Ferment
# William A. Shurcliff

The rudiments of the collection and storage systems employing water and air are illustrated in Figure 2.1. For simplicity, the controls and the equipment for delivering the heat to the rooms have been omitted. Typically, each square foot of the collector receives about 1,200 Btu of (0.3 to 3 micron) solar radiation per fair day in winter, and delivers about 600 Btu of that energy to the storage system. If a typical house in Massachusetts is fitted with a 500 square foot collector, about half of the heat that will be needed during winter can be supplied by the solar heating system; the other half is provided by a conventional furnace or electric heater.

Is there one clearly most promising avenue of solar collector design? Are there two or three clearly superior avenues? Unfortunately not. Perhaps there will be—in five years. But today, to the dismay of newcomers to the field, there are literally hundreds of approaches, and the race is neck-and-neck. No one knows which approach, or which dozen approaches, will eventually gain a decisive lead.

A vast search for "five percent improvements" is underway. If a few little improvements—in efficiency, in durability, in costcutting—can be made in schemes $P$, $D$, $Q$, those schemes may pull ahead of the rest. Any one of a hundred schemes, if slightly improved, may be among the winners:

The competition involves, mainly, petty details. No new principles of physics are expected, or sought. Rather, the emphasis is on salvaging a bit of otherwise lost energy here or there, reducing corrosion of a certain aluminum component, preventing a plastic sheet from warping or discoloring, eliminating a valve, finding cheaper materials and cheaper ways of fabricating them.

The main hurdles that the developers of active-type solar heating systems face are

- achieving high efficiency of solar-radiation collection,
- achieving high durability of equipment, a long life,
- arriving at a product that has very low cost, and
- devising equipment that can be successfully "retrofitted" to existing houses.

### Achieving High Efficiency

Typically, about half the radiant energy incident on a collector is successfully delivered to the storage system (or directly to the rooms). That is, the efficiency is about 50 percent. Not surprisingly, many investigators, es-

William A. Shurcliff

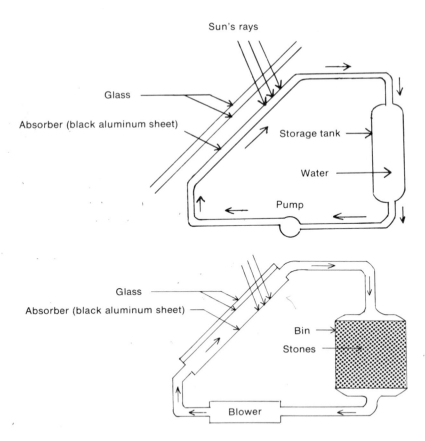

**Figure 2.1**
The flow of water or stream of air carries the heat from the solar radiation collector to the storage system (a water-filled tank or a bin filled with stones) or delivers it directly by conventional means (radiators or forced hot air) to rooms of the house. The upper diagram shows an active type of solar heating system employing water; the lower diagram shows an active type of solar heating system employing air.

Active-Type Solar Heating Systems for Houses

pecially those from prestigious universities and corporations, have spent years trying to increase the efficiency.

Many interesting advances have been made. But before reviewing these advances, we should ask: Is efficiency a clear-cut concept? And, is it really desirable to increase efficiency? Such questions have constituted a watershed. They have divided the investigators into high-technology and low-technology types; they have divided them into those that receive huge government grants and those that do not; they have divided them into those who have produced glamorous and highly uneconomic systems and those who have produced drab but economic systems. The two groups are scarcely on speaking terms. A silent battle is in progress. Presumably the two groups will pull together in the future as the relative merits of the two approaches become better understood.

*Is Efficiency a Clear-Cut Concept?* Considered naively, this question must be answered in the affirmative. Efficiency is a ratio of energies: solar energy captured and delivered for use in heating the house, divided by the total amount of solar energy incident. The definition is clear, and the quantity can be measured directly with the aid of pyranometers, thermometers, and flow-rate meters. The enormous literature on the subject has been summarized by Duffie and Beckman.[4] But there are catches, so many catches that some people wash their hands of the concept.
● The efficiency can exceed unity. If the outdoor temperature suddenly rises so high that it exceeds the temperature of the coolant entering the collector, the collector may gain more energy than is received as solar radiation: that is, the sum of the radiant energy absorbed and the in-leaks of energy via conduction and convection may exceed the incident radiant energy. To have an efficiency exceeding unity is disconcerting. It makes one suspect that there is something slightly wrong with the definition. (And something wrong with the goal? The goal is to heat the house, irrespective of where the heat comes from or how the losses arise.)
● The efficiency can be less than zero—it can be negative. If the collector is very hot, the outdoor air is cold, and the level of irradiation is very low, the collector, though accepting some solar radiation, may be losing more energy than it is gaining. Result: a negative efficiency. (One is reminded that there is a lot more to house heating than capturing radiation.)
● Instantaneous efficiency is of little interest. How the collector behaves at noon on the clearest day in winter is of little interest. What one wants to know is the performance throughout some longer period of time.

William A. Shurcliff

• There is no representative "longer period of time." Would it pay to concentrate attention on some one hour? No, because the hour chosen would not be typical of the day. Also, different hours (at different times of day or of year) lead to different directions of incident radiation, different intensities, different reflection losses from the collector glazing. Would it pay to concentrate on some one month? If so, which? December? February? May? How about considering the year as a whole: find that ratio of total energy captured throughout the year to the total solar energy incident? But this includes the summer, when one is already too hot and wishes not to collect energy!

• There is no one "standard condition of irradiation." Some collectors (single-glazed collectors, especially) perform well in mild months and poorly in the coldest months. Others (triple-glazed collectors) perform well in cold months but are disappointing in mild months. Some collectors (vertical ones) perform well when the sun is low in the sky, others (horizontal ones) perform well when the sun is high in the sky. How can collectors be compared meaningfully if there is no one standard condition of irradiation?

• There is no one "standard condition of collector coolant." Some collectors (for example, those of Harry E. Thomason[5] ) customarily receive coolant that is at 25 to 40° C and, not surprisingly, have small losses and high efficiency. Some other collectors customarily receive coolant at 60 to 90° C, hence "run hot," and may have much larger losses and lower efficiency, unless special precautions are taken.

Confronted with these awkward facts, some collector designers respond by providing not a single number representing the efficiency but a great set of graphs pertaining to a great variety of levels of irradiation, outdoor temperature, coolant inlet temperature, etc. But how is one to judge or use such a set of graphs? How are five sets of graphs, characterizing five kinds of collectors, to be compared? One reply is: "A weighting scheme is needed. One must evaluate a collector under all pertinent conditions, decide how often the various conditions prevail, and work out a weighted value of collection efficiency."

But this approach opens a new Pandora's box of troubles. The frequency, or prevalence, of a given set of conditions depends on many diverse and remote considerations. It depends, for example, on the temperature of the storage system, and this in turn depends on the size of the storage system, how well it is insulated, how well or badly the house itself is insulated, how often the owner's children leave windows open. Is the storage system allowed to get very cold in winter, or is it kept hot by an

Active-Type Solar Heating Systems for Houses

auxiliary heater? Also, what is the orientation of the house (and collector)? What is the tilt of the collector? How much sunny weather is expected at the location in question? What is the relative importance, here, of direct solar radiation and diffuse radiation? How intermittent is the radiation, typically? How much wind is there to cool the collector and reduce its efficiency?

In trying to characterize the collector, we must first characterize the location, the local weather, the storage system, the house, and the idiosyncracies of the occupants! The effective overall efficiency of a collector cannot be defined. So, how much time should one spend in trying to maximize something that cannot be defined?

If one must deal with efficiency, one should deal with the efficiency of the solar heating system as a whole. Or, better, the efficiency of the solar-plus-auxiliary heating system. Or, still better, the house as a whole. (Of course, if collector $A$ has a higher efficiency than $B$ under all conditions, the difficulties discussed above do not apply, and even the simplest comparison may be indicative and useful. But the important comparisons—the truly exciting ones—are comparisons between collectors of different types. When different types are involved, it will frequently be true that each type is superior under some conditions and inferior under other conditions.)

*How Can One Increase Efficiency?* Countless ways of increasing collector efficiency have been tried, often with success. Some of these ways are as follows:

● Use an electric-motor-powered tracking system to keep the collector aimed squarely at the sun.

● Apply an antireflection coating to the surface of the collector-window glass, thus reducing the reflection loss. Normally the loss is about 16 percent, and it can be reduced by a factor of about 5. Instead of using an antireflection coating, a special surface-etching process may be used.

● Use glass made of nonabsorbing ingredients (low-iron glass), thus reducing absorption loss (normally 4 to 10 percent) by a factor of 3 to 6.

● Instead of using glass, use a low-index fluorocarbon plastic, thus reducing reflection loss by a factor of nearly 2.

● Reduce conduction and convection losses by evacuating the space between the two layers of glazing (or between the glazing and the black absorbing sheet).

● Instead of using a vacuum, use an inert, high-molecular-weight gas; krypton, for example.

William A. Shurcliff

• To reduce convection between glass layers, or between glazing and black absorbing sheet, install a nonabsorbing egg-crate honeycomb here. This will let solar radiation through freely but virtually stop convective currents.
• To reduce the amount of energy carried away by the reradiation energy (wavelength 4 to 40 micrometers) from the hot, black, absorbing sheet, use a glazing material that has especially high absorption for such radiation. Or, apply to the glazing a special infrared-reflecting coating.
• To reduce the amount of energy radiated by the black absorbing sheet, apply a coating to it that—though strongly absorbing the (0.3 to 3 micrometer) solar radiation—has low emissivity in the reradiation band. More than 20 research groups have reported promising results with such selective surfaces.
• Instead of applying a special, low-emissivity coating, apply a micro-grooved structure to the black absorbing surface: a structure that discourages emission of long-wavelength radiation without discouraging absorption of solar radiation.
• To reduce the amount of energy radiated by the black absorber, keep the absorber cooler: that is, reduce the temperature difference ($\Delta T$) between the black absorber and the outdoor air.

To reduce $\Delta T$, replace the aluminum sheet with copper, use thicker copper, increase the number of coolant-carrying tubes in contact with the copper sheet, flow the coolant faster, arrange to keep the storage-system output-end cooler by making the storage system larger or by enhancing thermal stratification within it and drawing coolant from the coolest part.

To reduce losses at the collector edges, replace the all-metal frame with wood or with metal strips faced with an insulating material.
• Flank the collector with huge, crude, reflectors that will "funnel" additional solar radiation to it.

*Is It Really Desirable to Increase Efficiency?* The higher the efficiency, the better—other things being equal. But other things seldom are equal. Nearly every change made in the interest of efficiency increases the cost or reduces the reliability:

• Tracking systems capable of withstanding 100 mile per hour winds are very costly.
• Antireflection coatings are costly and, in most cases, not durable.
• Low-index fluorocarbon films may be expensive and fragile.
• Evacuating the space between glass layers entails many serious problems,

for example, resisting the crushing pressure of the atmosphere, maintaining high vacuum throughout many years. Also, vandals can easily cause spectacular implosions.

● Egg-crate honeycombs are costly, fragile and, if they become dirty, will absorb much solar radiation.

● Infrared-reflecting coatings may be costly and fragile.

● Microgrooved surfaces may be costly and may accumulate dust.

● Copper is more expensive than aluminum. If some copper and some aluminum are used, danger of harmful galvanic action arises.

● Enlarging the storage system, or enhancing thermal stratification within it, is expensive.

● Supplementary reflectors are expensive, may be ugly, may be damaged by high winds. The reflectivity will slowly deteriorate.

Each improvement in efficiency must be weighed against the increased cost and (often) decreased durability. Neither of these latter quantities is accurately predictable. Least of all is it predictable in advance of any routine production technique or long-term aging tests. Thus solar-heating system design is almost as much an art as a science.

The true goal of the solar-heating system designer is to devise a system that will collect much energy reliably and cheaply: the most energy per dollar. High efficiency per se is not a goal.

Experts versed only in engineering may not be the best persons to design solar heating systems—a truth that officials of the National Science Foundation (NSF) and the Energy Research and Development Administration (ERDA) and other agencies have been slow to learn.

### Achieving High Durability

The history of the development of solar heating is a history replete with tragedies: glass cracking, edge seals failing, tubes breaking loose from the metal sheet they had been soldered to, pipes becoming clogged, serpentine tubes failing to drain properly and subsequently freezing and bursting, joints leaking, insulation installed badly and allowing excessive heat loss, underground insulation becoming water-logged, rapid corrosion. Parts that hold up well in winter may deteriorate fast under the hot summer sun (and under glass!). In a four-day tour of twelve solar houses, the writer found that eight of them had broken glass and leaks of water or moisture. Mistakes are seldom publicized, however. Accordingly, newcomers to the field have frequently repeated the mistakes made earlier by others.

William A. Shurcliff

Veterans, especially those who have paid for their equipment out of their own pockets, have developed a passion for simplicity. They rank it high above elegance and efficiency. Some have turned to the use of passive collection systems, which are not discussed here, because of the extreme simplicity and reliability achieved there. Some veterans categorically refuse to use aluminum tubes. Despite various precautions, such as use of corrosion inhibitors, difficulties have been encountered with aluminum. Some other designers feel that aluminum must be used rather than copper because of its lower cost; they feel that control of corrosion must and will succeed.

Some designers have turned to the use of air-type collectors. Using air, one avoids a host of problems associated with water: for example, leaking, dripping, freezing, having to use conventional antifreeze (ethylene glycol), and having to interpose heat exchangers between the antifreeze and water used in the collector and the pure water used in the storage tank. But air is a poor medium in many ways: it has low density, low heat content per unit mass per degree of temperature; it is an insulator. A very high flow-rate is needed, and this may require a powerful, electric-motor-driven blower; and vibration and noise problems may arise. The large cross-section ducts required take up much room.

A question related to durability, or reliability, is this: Should the designer try to produce a solar heating system that will continue to function even if the supply of electricity from the local utility fails, for example, in the event of a local or widespread electrical blackout, such as might occur during wartime or during a stoppage of oil importation? Some solar heating systems are indeed independent of the electric supply; they are fully self-reliant, employing, for example, natural (gravity) convection (that is, thermo-syphon). But the degree of comfort provided, and the convenience, may suffer.

## Reducing Cost

At least 80 percent of the 200 or so solar heated buildings that exist today have been uneconomic. If the total cost (not only materials and labor but also time devoted to design, engineering, inspection, debugging, reparing, and the interest charges) of a typical system is computed, it far exceeds the benefits. More exactly, the total lifetime cost of building the equipment and operating it for, say, 20 years is much greater than the money saved through reduction in amount of fuel or electrical power used—

assuming that the costs of fuel and electrical power remain at today's levels. In most instances, the overall cost exceeds the overall benefit by a factor of 2 or 3. In some instances (where a high-technology firm worked against a 60-day deadline) the factor has been as high as 10.

**Warning:** The benefits of solar heating systems do not stem from fuel savings alone. The occupants of the house also benefit from the added security they have with respect to possible shortages of fuel or a huge rise in price of fuel. Also, they may find the solar heating system to be fun, like having a yacht.

Of what value is a so-called demonstration project that, finally, demonstrates a cost-vs.-benefit ratio of 3 to 1 or 10 to 1? Some cynics have gloomily surmised that some of these "demonstrations" were deliberately carried out so as to convince the public that solar heating is utterly impractical today and that the country should rush forward with other technologies, gies, such as nuclear. This author knows of no evidence to support this conjecture.

Solar heating veterans such as H. C. Hottel, M. Telkes, R. W. Bliss, H. R. Hay, S. Baer, H. E. Thomason, N. B. Saunders in the United States; A. E. Morgan in Great Britain; and F. Trombe in France have long realized the paramount importance of keeping the cost low. They have learned to quickly reject schemes that, however appealing technically, are incompatible with truly low cost. Many newcomers, too, gravitate toward this view. Many others are still preoccupied with technical elegance, blindly trusting that, somehow, cost problems will vanish when mass production is begun.

Is this trust well founded? Will mass production drastically reduce cost? Costs can certainly be reduced below today's levels—assembly costs, especially. But whether the reduction will be substantial or only moderate is uncertain. Why? Because (a) components such as glass, sheet metal, tubes, insulation, and valves are already being mass produced; and (b) the equipment is bulky and fragile—shipping it (in crates?) to a site will always be expensive, and installation work will often be "special" and may remain expensive. Electric clocks and transistor radios are truly cheap, but houses and solar heating systems are a different kettle of fish. Consider the cost of modernizing a bathroom: the new equipment may cost only $300, but the total bill may reach $3,000.

### Retrofit Problem

Today, most solar heating systems have been applied to new houses that

William A. Shurcliff

are superbly insulated and are situated in rural or suburban locations where there is unobstructed sunlight even in December. How much more difficult it is to "retrofit" solar heating to an existing, badly insulated, imperfectly oriented house in a region crowded with tall trees or tall neighboring buildings! Nearly any remodeling job tends to be surprisingly expensive, and to "remodel" a roof, "remodel" the basement, and install piping and controls may be correspondingly expensive. Yet for every house that will be built this year, there are 50 or 100 houses already standing. To solar heat new buildings will be helpful to our economy; but to solar heat existing buildings would be orders of magnitude more helpful.

Within three years, I believe, many truly economical schemes for solar heating new homes will be available; indeed, there are a few such schemes available today. But I am far from confident that—even in six years— economical systems applicable to the majority of existing houses will be available. I expect that, by that time, economical schemes will be available for the most favorable 25-percentile of those houses.

A solution to the problem of where energy can be stored in an existing house may be at hand. M. Telkes, in 30 years of tenacious research, has finally perfected a storage system that is 6 to 8 times more compact than a tank-of-water storage system of equivalent effective thermal content. The store employs a eutectic salt (sodium sulfate dekahydrate or sodium thiosulfate pentahydrate) that melts when warmed and solidifies when cooled— and gives out, or takes in, enormous quantities of heat (about 60 calories per gram) in undergoing such phase change. Many practical difficulties were uncovered and early trial use of the material was disappointing. But each difficulty has been overcome and, in mid-1975, exhaustive and realistic durability tests were completed successfully.[6]

## Other Problems

Besides the problems of performance, durability, and cost discussed above, there are other collateral problems.

*Conforming to NBS-HUD Standards* The National Bureau of Standards, on behalf of the Department of Housing and Urban Development, has already proposed standards for solar heating systems.[7] Some of the wording is highly technical and frightening. In some instances the wording is very vague, and this too can be frightening. Costly test instruments are called for as well as elaborate procedures.

Why have standards? The usual answer is: "To protect the public. To insure that it will get a high performance, durable product." But this author has grave misgivings. Preparation of standards is premature—if the standards are to be mandatory. If the government had set standards for automobiles in 1920, insisting, say, on self-starters, shatterproof glass, and 30,000-mile tires, every manufacturer would have had to shut up shop. Granted: purchasers of those cars suffered considerably in hand-cranking the cars; broken glass caused many injuries; tires seldom lasted 5,000 miles. Yet the benefits of the automobile were great, and the owners found the benefits to far outweigh the costs and difficulties.

Some cynics wonder whether or not the purpose of the proposed standards is to frighten away small producers of solar heating equipment and leave the field clear for the big corporations. It may not be too early for government-sponsored guidelines, but it is much too early to lay down elaborate and mandatory standards.

Solar heating is a great venture. The venture will falter if our government decrees that homeowners must take no risk. Life today is fraught with risks. Why must solar heating be kept free of risk? And is there not a greater risk: that the house will be stone-cold, some winter?

*Living with a Secrecy and Patents Roadblock* The race to improve solar heating systems is impeded by secrecy and patents. Already there are hundreds of issued patents (most of them dating back 10 to 100 years), and presumably there are hundreds of patent applications in progress. Many patents are vaguely drawn. Many seem near duplicates of others. Many seem of doubtful novelty, hence doubtful validity. There is no published comprehensive book on solar heating of houses. Old ideas are reinvented by newcomers. Many persons feel that others are trespassing on their patent rights. Hard feelings arise. People tend to be secretive about their developments. Should a manufacturer embark on a large production program while uncertain as to whether his design will prove to be covered by another's patent?

*Avoiding Being Mesmerized by the Lure of Government Grants* Repeatedly, the NSF or ERDA or other agencies have offered to make grants to institutions, corporations, or persons hoping to develop better solar heating schemes. Thousands of applications for grants, each a major literary undertaking in its own right, are submitted by able people, who then wait for

William A. Shurcliff

half a year only to learn, finally, that their proposals are rejected. There are few winners. Thousands of man-hours are spent by highly competent engineers in preparing the proposals; sometimes the cost of the mass of proposals nearly equals the total amount of money in the set of grants.

The number of government grants is modest, and most of the grants have been to big name universities and corporations. To give a grant to a lone-wolf inventor with nothing in his favor (except perhaps a truly successful solar-heated house!) takes more courage than government agencies can muster. Perhaps the situation is improving. (Perhaps, also, individual states should give support to solar heating projects tailored to local or regional conditions.)

*Building for Tomorrow, Not Today* Whether a given solar heating system is economic today is scarcely of interest. It is tomorrow that counts. How fast will the prices of oil, coal, electricity rise in the next few years? Will costly solar heating systems such as are being built today (uneconomically) be economic then? Can a manufacturer stake his future on a 100 percent rise in cost of conventional fuel? Alas, there are no good answers to these questions.

*Arranging State-Sponsored Monetary Inducements* Many states have enacted legislation aimed at inducing home-owners to buy solar heating equipment. The more systems installed, the less our dependence on foreign sources of oil and the better our balance of payments. The inducements proposed have been low-interest loans for purchase of solar heating equipment, and tax reductions based on the amount spent for such equipment.

*Other Questions* Companies proposing to sell active-type solar heating equipment are faced with other questions. For example, should designers aim for solar heating systems that can provide 100 percent of the winter's heatneed? Or should they settle for 50 percent? Most experts believe that 50 to 70 percent solar heating is optimum, throughout most of our country. To achieve 100 percent might require, say, a collector three times as large and expensive. (Yet the added security of a 100 percent system, and the sport of it, might induce some owners to shoot for 100 percent.)

Will passive-type solar heating, with its extreme simplicity and low cost, preempt the field? Must a solar heating system, to be truly economic, be an integral part of the house? Or can it succeed as a bolt-it-on appliance?

Will heat pumps of improved type (higher coefficient of performance) take the center of the stage? Will someone work out a successful scheme that uses heat pumps in conjunction with solar collectors—the collector operating at low temperature with extremely high efficiency, and the heat-pump "upgrading" the energy for storage in a small but very hot tank?

Will there be, ultimately, one or two winning schemes, or will there be dozens, each appropriate to a different climatic region, different size or shape of house, or different degree of affluence of the occupants?

What about large commercial buildings? Some persons favor concentrating on these because (a) such buildings have large volume-to-surface ratio, hence require less heat per unit of volume or per unit of floor area, that is, they are easier to heat; (b) they have large, easily accessible roof areas and installing collectors there is easy; and (c) disfiguring a commercial building seems less reprehensible than disfiguring a house. Already, a dozen large commercial buildings have been equipped with solar heating.

Can solar heating of houses stand on its own feet or should solar cooling be tied to it, so that the collector can be used summer and winter, reducing the amortization charges against the winter use? For large buildings in the south, some tie-in may be appropriate, but the subject is highly technical and is outside the scope of this essay.

Is the solar heating of houses closely related to the solar heating of water for household domestic hot water supplies? Probably not. For domestic hot water heating, a small (8 ft × 4 ft, or 8 ft × 8 ft) collector usually suffices, and such a collector can indeed be factory built, as an appliance, and bolted onto the house roof. Being used throughout the year, such devices can be especially cost effective. Already there are about 100 firms making solar hot water heaters in this country, and it is rumored that there are of the order of 100,000 such devices already in use. Such devices are outstandingly popular in Israel, Japan, and Australia.

## Summary

There are thousands of ways of building a solar heated house. Hundreds of ways are being tried out. Scores of very different ways have their individual special merits. Some are good for cold climates, some for hot. Some are favored by very rich people, others by very poor. Some provide great convenience; others are much cheaper but cruder. Some will last 20 years, others will need reconditioning in 5 years. Some wreck the design of the house; others don't. Some are well proven; some are not yet proven but

William A. Shurcliff

may be much cheaper. Some you can buy; others you must build yourself. Some fit in with cooling schemes, greenhouse schemes; others do not. We deal, indeed, with a ferment.

A majority of the solar heating systems in use today are hopelessly uneconomic. A few are truly economic, but have been little noticed by big industry and government. Contrary to the prevailing view of ERDA and related agencies—that the immediate task is to get mass production underway—the first task is to identify, or invent, schemes that have real prospect of overall success. We are still in the romantic, wide open, inventing stage. Thousands of eager people, most of them young, are involved. Their purses may be near-empty; but their smiles are infectious. They, and this author, are highly optimistic.

## Notes

1. W. A. Shurcliff, *Solar Heated Buildings: A Brief Survey* (13th rev. ed.: 19 Appleton St., Cambridge, Mass. 02138: The author, 1977).

2. Shurcliff, "Informal Directory of the Organizations and People Involved in the Solar Heating of Buildings" (3rd ed.: Cambridge, Mass.: The author, 1977); U.S. Energy Research and Development Administration, "Catalog on Solar Energy Heating and Cooling Products," Document ERDA–75 (Washington, D.C.: ERDA, 1975).

3. See, for example, F. Daniels, "Direct Use of the Sun's Energy" (New Haven: Yale University Press, 1964); F. de Winter, *Solar Energy and the Flat-Plate Collector. An Annotated Bibliography* (New York: Copper Development Association, Inc., 1974); A. R. Patton, *Solar Energy for Heating and Cooling of Buildings* (Park Ridge, N. J.: Noves Data Corp., 1975); S. V. Szokolay, *Solar Energy and Building* (London: Architectural Press, 1975); and J. A. Duffie and W. A. Beckman, *Solar Energy Thermal Processes* (New York: John Wiley & Sons, 1974).

4. Duffie and Beckman, *Solar Energy*.

5. H. E. Thomason and H. J. L. Thomason, *Solar House Plans II-A* (Barrington, N.J.: Edmund Scientific Co., 1975).

6. M. Telkes, "Thermal Storage in Sodium Thiosulfide Pentahydrate," paper presented at Meeting of IECEC, Aug. 18, 1975, Univ. of Delaware.

7. National Bureau of Standards "Interim Performance Criteria for Solar Heating and the Combined Heating/Cooling Systems and Dwellings" (Washington, D.C.: Department of Housing and Urban Development, Jan. 1, 1975).

He approached the problem thus: "When one means to have the right sort of house, must he contrive to make it as pleasant to live in and as useful as can be?"

And this being admitted, "Is it pleasant," he asked, "to have it cool in summer and warm in winter?"

And when they agreed with this also, "Now in houses with a south aspect, the sun's rays penetrate into the porticos in winter, but in summer the path of the sun is right over our heads and above the roof, so that there is shade. If, then, this is the best arrangement, we should build the south side loftier to get the winter sun and the north side lower to keep out the cold winds. To put it shortly, the house in which the owner can find a pleasant retreat at all seasons and store his belongings safely is presumably at once the pleasantest and the most beautiful."

Xenophon<br>
Memorabilia Socratis<br>
III:viii

The comfort of a house and its heating fuel requirements, too, depend greatly upon the way the house is built. To a considerable extent, the heating fuel requirements of a house are essentially built into it at the time of construction and cannot easily be changed very much after it is built. If a house is so built that it wastes fuel it will probably continue to do so, and will waste a very large amount of fuel during its long lifetime.

Energy is supplied to the interior of a house from a variety of sources, and all this energy goes somewhere. Insofar as house-heating is concerned, the balance between energy gains and losses may conveniently be separated into the energy balance equation:

GAINS = LOSSES,

$$G_1 + G_2 + G_3 + G_4 = L_1 + L_2 + L_3,$$

where

$G_1$ = solar energy in through windows

$G_2$ = All electrical energy consumed in house, except that used to heat domestic hot water

$G_3$ = thermal energy given off by occupants

$G_4$ = "fuel" energy supplied to interior of house by house-heating system

$L_1$ = energy loss out through windows

$L_2$ = loss through roof, walls, doors, and through floor (or through cellar)

$L_3$ = energy required to heat outdoor air which leaks through house.

---

# 3 Why Not Just Build the House Right in the First Place?
# Raymond W. Bliss

Often the "fuel" energy supplied to the interior of a house by its heating system—$G_4$ in the energy balance equation—does come from the burning of an actual fuel such as oil or gas. In other houses it may come from electric resistance heating, from a heat pump, or perhaps from a solar-collector system which converts solar energy to thermal energy outside the house and brings it in. Present prices being what they are, most people take a dim view of the fuel energy cost, $G_4$, in the energy balance equation. They would prefer that $G_4$, especially in their own house, be somewhat smaller than it is, so that they would not have to pay as large a heating bill.

Obviously, $G_4$ can be reduced by reducing any of the losses, by increasing any of the other gains, or by some combination of both. Essentially, this essay is a discussion of what can be done practically in new house construction to reduce the amount of fuel energy needed ($G_4$) by reducing the loss of energy ($L_2$) through the roof, walls, etc., by reducing the amount of energy needed to heat outdoor air that leaks into the house ($L_3$), and by increasing the solar gain through windows ($G_1$). Again, rather obviously, $L_2$ can be reduced by more thorough insulation at appropriate places, and $L_3$ can be reduced by building the house tighter, so as to cut down on air infiltration.

In a typical house the majority of air infiltration is associated with cracks around windows and doors. The remainder—almost always, a significant fraction of the total—is associated with a variety of other cracks, holes, vents, porous walls, etc. The solar gain through windows, $G_1$, can be increased by using larger windows or, more importantly, by placing the maximum practicable fraction of total window area at the orientation receiving maximum sunshine (that is, south). My intention here is to "put some numbers" to the above concepts.

American houses, as presently built, are not really flagrant fuel wasters, though some office buildings are. As a fuel-waster, a currently-built house should not be likened to the 8-miles-per-gallon cars which were built a few years ago. The fuel economy of currently-built houses might better be likened to that of a 15-mile-per-gallon car: that is, its fuel economy is not too bad, but is readily susceptible of improvement.

Somewhat analogous to improving the fuel mileage of a car by making it smaller or less powerful, the heating fuel requirements of a house can be reduced by making the house smaller or maintaining it at a lower indoor temperature. Alternatively, in the case of a car, its fuel requirements can also be lessened by building it somewhat better, at somewhat greater cost, but retaining the same size and power. The same thing is true of a house,

Raymond W. Bliss

and to a greater degree: it is readily possible to design and build a house just as large as conventional construction, and keep it just as warm, but which requires only half as much heating fuel as current construction. And, most importantly, it is readily possible to do this at an increase in construction cost which is small enough to be repaid reasonably promptly in fuel savings.

Although the design of houses for fuel saving is an important matter, one must always keep in mind that houses are designed primarily for living; and man does not live by saving Btu's alone. An engineer with a narrow specialty (this writer is an engineer with the narrow specialty of heat transfer) must constantly remind himself that any fuel-saving design improvements he may conceive of for a house must do more than merely fit his own ingrained desire for "cost-effective efficiency": more importantly, those design improvements must contribute to a house that most people would consider pleasant to live in, comfortable and beautiful.

One has to start somewhere though. In this essay fuel-saving design is considered primarily from the narrow standpoint of an engineer. Both winter-heating and summer-cooling energy requirements are, of course, important matters in house design; but here, for reasons of brevity, I will focus on the winter-heating aspects.

## Fuel-Saving Principles

The basic principles of designing a house for winter fuel-saving in most northern U.S. climates are simple enough.

1. Choose window placement and size so as to get as much sunshine as possible into the house in winter and keep as much as possible out in summer.
2. Reduce shell heat loss (heat loss through roof, walls, doors, windows, and floor) to best economically practicable extent.
3. Reduce air infiltration to best economically practicable extent.

Application of the above basic principles involves a great many details relating to climate, to the shape and orientation of the house, to energy gains and losses of building components, etc. Few of these details are overly complicated. No one of them, by itself, is of outstanding significance; many of them are reasonably important. This makes it difficult, in any discussion of fuel-saving houses, to steer a suitable course between the Scylla of meaningless generalities and the Charybdis of interminable descriptions of seemingly trivial details.

The course adopted here is to choose a particular location and climate, assume several specific similar houses placed in that climate, present their calculated seasonal heating energy requirements, and discuss the matter as we go along. The particular location chosen is Boston, Mass. The average sunshine and temperature conditions for Boston are shown in figure 3.1. (As far as I am aware, average weather conditions have never yet actually occurred in Boston but, during some future winter, they might.) The sunshine data shown in figure 3.1 are from the measurements of Charles Cuniff;[1] the temperature curve is from U.S. Weather Bureau data.[2]

As one example of a specific house I chose the very conventional two-story house shown in figure 3.2, with two variations in window placement and two variations in heat loss rate parameters. A total window area of 300 square feet is assumed, and the two assumed window placements are shown in table 3.1. It is seen that the difference in the two window placings amounts merely to moving 80 square feet of window from the north to the south side of the house.

For heat loss rate parameters I assumed the two sets—FHA and Better—shown in table 3.2. The FHA parameters represent, essentially, the worst the Federal Housing Administration will put up with and still grant that house an FHA-approved mortgage.[3] As usually happens with minimum standards of this sort set up by government regulatory agencies, the concerned industry tends to adopt them as a ceiling rather than a floor on quality. Builders seldom build houses with lower heat loss rates than the highest the FHA will put up with. The better parameters listed in table 3.2 illustrate somewhat better construction that can easily be built, costs a little more but would, I believe, pay for itself reasonably promptly in fuel saving. (FHA regulations relating to maximum allowable heat losses are not actually as simple and explicit as might be inferred from table 3.2. Also, they are changed from time to time. Table 3.2 gives a reasonably close approximation of current FHA regulations for climates similar to Boston.[3] In general, current FHA regulations are considerably stricter than those of a few years ago. Readers interested in the exact details of FHA regulations should, of course, consult up-to-date copies of the regulations themselves.)

In using the loss rate parameters of table 3.2 to calculate the energy loss of a house, no safety factor for insulated components was assumed. That is, it was assumed that the actual loss rate through insulated components of the house would conform with handbook values. Insulation carefully and competently installed in an actual house does have actual loss rates in close agreement with handbook values. If carelessly installed, it can have

Raymond W. Bliss

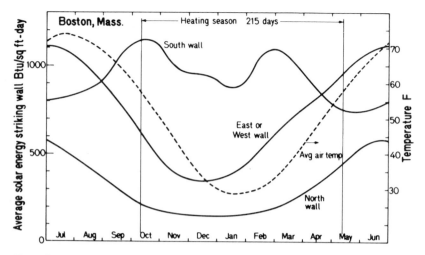

**Figure 3.1**
Sunshine and temperature averages for Boston, Mass. Data assumed to calculate seasonal heating energy requirements for houses in Boston; average solar energy striking wall of the house and average temperatures. *Source:* For sunshine data, see Cuniff[1]; and for temperature data, see U.S. Weather Bureau[2].

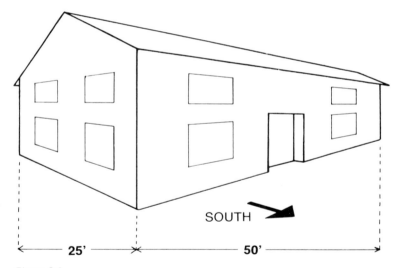

**Figure 3.2**
The assumed conventional house. A two-story house; total window area of 300 square feet; average occupancy of four persons; and maintained at 70° F at all times is used to calculate energy requirements during average heating season in Boston.

Why Not Just Build the House Right?

**Table 3.1**
Window placement in conventional house. "Better than standard" is achieved by merely moving 80 square feet of window from north to south side.

| | Window area on each wall (square feet) | | | |
| --- | --- | --- | --- | --- |
| | S | N | E | W |
| "Standard" window placement | 100 | 100 | 50 | 50 |
| "Better" window placement | 180 | 20 | 50 | 50 |

**Table 3.2**
Energy loss rate parameters. Data and assumptions assumed to calculate energy loss rates for conventional house.

| | FHA | Better |
| --- | --- | --- |
| Roof | 0.053 (Frame construction 7 inches insulation) | 0.053 (Same as FHA) |
| Walls | 0.085 (Frame construction 3 inches insulation) | 0.065 (Frame construction 5 inches insulation) |
| Doors | 0.65 (Ordinary door) | 0.30 (Ordinary door plus storm door) |
| Floors[a] | 0.084 (Perimeter insulation 1 inch thick and 18 inches deep) | 0.025 (Perimeter insulation 4 inches thick and 48 inches deep) |
| Windows | 0.65 (Double-glazing with 3/16 inch air space, or single-glazing plus poorly-fitted storm window) | 0.54 (Double-glazing with 5/8 inch air space, or single-glazing plus tight-fitted storm window) |
| Air infiltration (in changes per hour) | 1.0 | 0.5 |

Note: FHA means minimum construction requirements for an FHA-approved mortgage: a ceiling rather than a floor on quality. The numbers are "U-values," apparent overall thermal conductances in Btu per °F per square foot per hour. Examples of typical construction that give approximately these U-values are in parentheses.
[a]The U-value for floors is given here on a per square foot of floor basis; and the FHA-value (0.084) pertains to a 50 × 25 foot "slab-on-grade-floor."

Raymond W. Bliss

much higher loss rates. Electrical heat gains were assumed as about 20 kilowatt hours per day; occupancy was assumed to be four persons. All rooms of the house were assumed to be maintained at 70° F at all times. Using all the preceding data and assumptions, the seasonal energy balances for the four variations of the assumed conventional house were then calculated, and the results are presented in figure 3.3.

The first energy balance of figure 3.3 refers to the conventional house with standard window placing and FHA loss parameters. Its seasonal energy requirement from the heating system is 92 million British thermal units (Btu). Since this version of the house is typical in every way, and would just meet FHA standards on heat losses for a house of this size, shape and window area, it is used as a sort of reference standard against which to compare the heating energy requirements of other versions of this particular house. The heating energy requirement of our reference standard is referred to as 100 percent of FHA (see fig. 3.3–A).

The next version of the house is shown by the second energy balance of figure 3.3, which illustrates the effect of shifting 80 square feet of window from the north to the south side of the house, with no other change in house design. The window shift results in an additional solar gain of 9 million Btu, which brings the heating energy requirement down by the same amount, that is, down to 83 million Btu from the 92 million Btu of our reference standard. Since 83/92 = 0.90 approximately, the heating energy requirement of this version is referred to as 90 percent of FHA (see fig. 3.3–B). The saving of 9 million Btu produced by this window shift is not overly spectacular. But it is to be noted that this particular energy saving adds nothing to construction costs–it costs no more to build a window in the right place than in the wrong place.

The economic value of any energy-saving procedure depends upon the cost of the energy being saved. In this connection, there are two convenient and easy-to-remember approximations:

• If a house is heated with oil costing 40 cents per gallon, the fuel costs of delivering a million Btu to the interior of the house is about $4.
• If a house is resistance-heated with electricity costing 4 cents per kilowatt hour, the fuel cost of delivering a million Btu to the interior of the house is about $12.

Behind the $4 per million delivered Btu for oil are two unstated assumptions: (1) At the furnace, 70 percent of the energy contained in the oil is delivered to the heating system, and the remaining 30 percent goes up the chimney. (2) All of the 70 percent delivered by the furnace goes to rooms which it is desired to heat. The first assumption is quite valid for a good-quality residential furnace, provided it is well-adjusted and well-maintained. The second assumption is often not so valid, notably so in cases where the furnace is located in a room that the homeowner does not desire to heat (typically the basement) and in cases where the energy lost from ducts or piping of the heating system leaks directly outdoors without contributing to heating the rooms that are to be heated. In such cases the fuel cost of a million Btu actually delivered to the rooms which the homeowner wishes to heat can be considerably more than the $4 mentioned.

Using the preceding approximations, the yearly dollar saving to the home owner resulting from building those 80 square feet of window on the south side instead of the north side, and thereby saving 9 million Btu per year, would be about $36 per year if he heats with 40 cents per gallon oil and about $108 per year if he uses resistance heating with electricity costs at 4 cents per kilowatt hour.

The next energy balance of figure 3.3 illustrates a more spectacular saving—that attainable by using the "better" loss parameters of table 3.2, with window placing in standard position. This saves 49 million Btu, bringing seasonal heating energy requirement down from 92 to 43 million Btu, which is 47 percent of FHA (see fig. 3.3–C). This 49 million Btu saving did not come free, however. It cost extra—probably about $1,500—to make all the improvements in loss rate parameters from the FHA to the better values of table 3.2. If it did cost $1,500, then the cost/benefit ratio associated with making all these improvements can be expressed as $1,500/49 or a $31 initial cost per yearly million Btu saving.

If a house is heated with oil or other fuel costing $4 per million delivered Btu, a $31 initial investment to save a million Btu yearly would be repaid in about 8 years in fuel savings. Similarly, if a house is heated with resistance heating or any other fuel costing $12 per million delivered Btu, then the same $31 initial fuel-saving investment would be repaid in less than 3 years. A substantial percentage of new U.S. housing, incidentally, is electrically heated, and the percentage has been increasing each year. In 1972 about 36 percent of the homes built that year were electrically-heated, 54 percent gas-heated, and 8 percent oil-heated.[4]

The last balance of figure 3.3 illustrates the effect of combining both improvements—better loss rate parameters and better window placing. This saves an additional 7 million Btu compared to the previous balance, bring-

Raymond W. Bliss

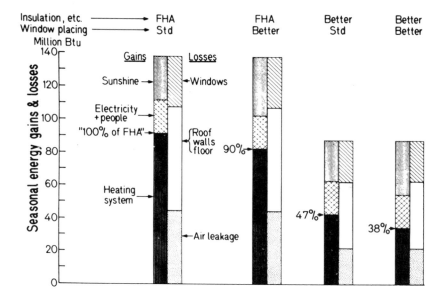

Figure 3.3
Four variations in conventional house. Calculated heating energy requirements for variations in window placement and allowable heat losses: (A) standard window placement and standard FHA allowable heat losses; (B) 80 square feet of window shifted from north to south side of house; (C) better than standard allowable heat losses with standard window placement; and (D) better window placement and better than standard allowable heat losses.

ing heating energy requirements down to 35 million Btu, which is 38 percent of FHA (see fig. 3.3–D).

It will be noted that shifting the 80 square feet of window from north to south saved 7 million Btu in this case, whereas the same shift saved 9 million Btu in the previous case of the less well insulated house. The same amount of sunshine came through the windows in each case (the crude houses postulated for the calculations lying behind figure 3.3 do not include any consideration of overhangs to shade the windows). The reason for the 2 million Btu discrepancy is that the better-insulated house of this case received 2 million Btu more than it actually needed during early fall and late spring. This unusable excess would tend to overheat the house and must be discarded. In a well-designed house most of the excess is pre-

Why Not Just Build the House Right?

cluded from entering the house at all, by properly-designed window-shading overhangs at the south wall. Any remaining excess that overheated the house would be discarded, naturally, by the homeowner by pulling curtains, opening windows and doors, etc.

The reference standard house of figure 3.3 has a heating energy requirement of 92 million Btu; the best energy-saver of the group—the last balance of figure 3.3—requires 35 million Btu, which is a saving of 57 million Btu. It is instructive to see just how this 57 million Btu was saved, so we have tabulated the various savings in table 3.3 (in which, due to rounding, the savings total is 56 million Btu). It is seen that the largest savings are those associated with cutting down air infiltration and reducing floor losses.

Roughly similar savings can be had with any new conventional house. They can easily be built into it by any competent and reputable builder provided he puts his mind to it and has a buyer agreeable to paying the reasonable extra construction cost. Every one of the 1.5 million or so houses that will be built in the United States during 1976 could be built, at very reasonable extra cost, to have an energy requirement of 50 percent of FHA. They will not actually be so built, however. A goodly percentage of them will have heating energy requirements in the vicinity of 100 percent of FHA. Most of the rest of them will have heating energy requirements greater than 100 percent of FHA.

If the heating energy requirement of any ordinary house can be reduced so easily to 50 percent of FHA, how far can one go in reducing heating energy consumption by custom design? The answer, roughly, is that one can get down to around 20 percent of FHA by custom design—but there are some complications.

### Directly Solar Heated

In theory, one could achieve 20 percent of FHA, or even zero percent of FHA, by using exorbitant amounts of insulation (say two or three feet thick) and building the house so tight that ventilation was controlled to the minimum required for breathing. That approach is neither economically nor aesthetically practical. The more practical approach, at least in most of the northern parts of the United States, is to use more sunshine. This leads to a house design which we call a directly solar-heated house.

By a directly solar-heated house we mean, simply, a very well-insulated house (by current standards), with large south-facing windows and a minimum of window area at other orientations. As a descriptive term, directly

Raymond W. Bliss

Table 3.3

Summary of energy savings (56 million Btu saved by going from standard FHA allow-able heat losses and standard window placement to better than standard allowable heat losses and better window placement)

| Item | Seasonal saving (million BTU) | Method |
|------|-------------------------------|--------|
| Less air infiltration | 22 | Infiltration reduced from 1 air change per hour to one-half air change per hour, by tighter construction |
| Less floor loss | 12 | Better perimeter insulation (see table 3.2) |
| Less wall loss | 7 | Wall U-value changed from 0.085 to 0.065 (see table 3.2) |
| Increased solar gain | 7 | 80 square feet of window shifted from north to south side of house |
| Less window loss | 5 | Window U-value changed from 0.65 to 0.54 (see table 3.2) |
| Less door loss | 3 | Storm doors at all exits (3) |

solar-heated is not ideal, because any house with windows has some solar gain and is, at least to some extent, directly solar-heated. (Government agencies involved in the federal solar energy programs divide solar house-heating methods into two general types: "active" types, which require machinery for their operation; and "passive" types, which do not require machinery. What I call "direct solar-heating" would be called "passive solar-heating" in government terminology.)

However, from a designer's standpoint, there is a rather sharp demarcation between an ordinary house—moderately insulated and with moderate window area—and a directly solar-heated house—quite thoroughly insulated and with relatively large south-facing windows. An ordinary house, no matter how it is designed, will not overheat in winter due to solar gains through the windows (although it may well do so in summer). On the other hand, a directly solar-heated house, with its combination of unusually good insulation and unusually large solar gain, must be carefully designed to prevent overheating on clear days during the winter heating season.

There are two basic design requirements to prevent such overheating:

• Adequate thermal storage capacity on the indoor side of the house insulation must be provided to "soak up" excess mid-day solar gains for later use.

Why Not Just Build the House Right?

● This storage capacity must be so arranged that it soaks up the incoming solar energy fast enough—in other words, the excess solar energy coming in the windows at mid-day must be transferred to storage just about as fast as it comes in.

One good way to meet these two requirements is to design the house with walls and floor of heavy construction (masonry, stone, or concrete) with enough total weight to meet the first requirement and with suitable placing, thickness, surface areas, etc., to meet the second requirement. Heavy construction of this type seems practically mandatory for a directly solar-heated house designed for particularly low heating energy requirements, for example, in the vicinity of 20 percent of FHA.

Also, of course, it is especially important to keep unwanted summer sun from entering the large south-facing windows of a directly solar-heated house. For such a house, window overhangs and other shading devices must be accurately tailored to the climate, latitude, and house in question.

It is because of these design problems that a directly solar-heated house must be custom-designed. The greater the relative solar gain desired, the more care must be given to the design. A directly solar-heated house designed to have a heating energy requirement of 20 percent of FHA requires very extensive design work and close collaboration between architect and engineer.

Acceptably uniform interior temperatures are most conveniently attained in a directly solar-heated house if its interior plan is somewhat more open (that is, somewhat less cut up into small rooms) than is customary in most U.S. houses. The optimum shape of a directly solar-heated house is of course strongly governed by the path of the sun, and its path tends to dictate a shape quite similar to that recommended by Socrates 24 centuries ago.

Figure 3.4 illustrates a suitable configuration for a custom-designed "20 percent of FHA house." It is a single-story high-ceiling house, with south side "loftier to get the winter sun," and with a total window area of 420 square feet, of which 360 square feet are on the south side. We have calculated the seasonal heating energy requirements of this house, on the same basis as described for the conventional house already discussed, for two variations:

● House turned 180 degrees, so that the large windows face north and with FHA loss parameters.
● House facing south with "thorough" insulation. Thorough insulation

Raymond W. Bliss

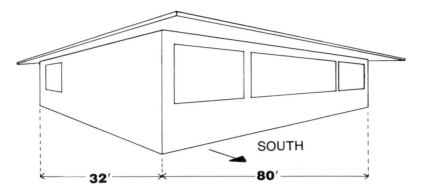

**Figure 3.4**
Custom-designed house. A one-story high-ceiling house with south side "loftier" to get the winter sun; total window area 420 square feet, with 360 feet facing south; heating energy requirement would only be 20 percent of that needed for standard FHA conventional house of same size and shape (assumed location, Boston, Mass.).

means the same loss rate parameters as those of the better column of table 3.2, except that walls are assumed insulated to achieve a value of $U = 0.05$ (as in the roof) and insulating panels are used on all windows at night.

The resulting energy balances are shown in figure 3.5, in which the bars have the same meaning as those of figure 3.3. The north-facing FHA insulated version is seen to have a seasonal heating energy requirement of 123 million Btu; the south-facing thoroughly insulated version has a heating energy requirement of 21 million Btu. The useful solar gain of this version is large, about 49 million Btu, as can be seen from figure 3.5. The total solar gain during heating season (with no shading overhangs) would be about 56 million Btu.

The heating energy requirement of the south-facing thoroughly-insulated version is about 17 percent of FHA if the north-facing FHA-insulated version is taken as a reference standard. Since most people do not build their houses with practically all windows facing north, a more logical comparison of heating energy requirements is that between (1) the south-facing thoroughly-insulated version and (2) an ordinary single-story house of the same floor dimensions, with windows uniformly distributed over wall area and FHA loss parameters. The calculations for that comparison were done too, and compared on that basis the south-facing version is about 20 percent of FHA.

Incubator-like precision of indoor temperature control is not feasible in a house with large south-facing windows. Winter indoor temperatures in a

Why Not Just Build the House Right?

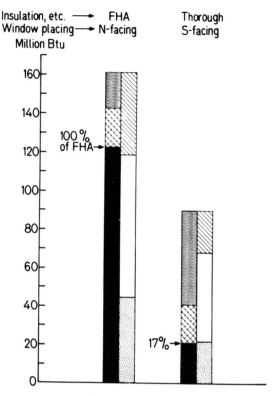

Figure 3.5
Two variations in custom-designed house. Projected heating energy requirements given variations in window placement and allowable heat losses: (A) large windows facing north and standard FHA allowable heat losses; and (B) large windows facing south and "thorough" insulation.

Raymond W. Bliss

well-designed 20 percent of FHA house probably would fluctuate between 65° F and 85° F. Summer indoor temperatures, in the Boston climate, probably would fluctuate within about the same range.

The construction cost of a directly solar-heated house is on a par with other custom-built construction, and the overall practicality of a directly solar-heated house is really more a matter of personal preferences than of economics. A directly solar-heated house has certain inherent features which appeal to some people and not to others. The shape and orientation of such a house are inherently dictated by the path of the sun, and tend to be similar to that of figure 3.4. The southern rooms of a directly solar-heated house are very brightly lighted on sunny days, objectionably so to some people. The broad window expanses tend to make one acutely aware of the ever-changing outdoor weather. The indoor temperature fluctuates. Some people like this sort of thing, others do not. For those who like the inherent characteristics of a directly solar-heated house, and have an appropriately oriented site on which to place it, a directly solar-heated house can be most pleasant and can save dramatic amounts of fuel.

The vast majority of new U.S. housing is not custom-designed. Few new homes are owner-designed and built; even fewer, probably, are architect-designed. Consequently, at present, the more significant opportunities for large-scale fuel-saving in house construction lie in the great yearly flood of conventional housing construction built by regular builders. From a national energy conserving standpoint, enormous fuel savings would be possible if this yearly flood of conventional construction could be built with an eye to better fuel-saving.

For the conventional house in Boston used as an example in these notes, it appeared that point-of-use energy savings of around 50 million Btu per year could be attained by simple, reasonably inexpensive modifications usable by any builder—better window placing, better insulation at various places, tighter construction to reduce air infiltration, etc. The conventional house of our example has an inside floor area of around 2,350 square feet, which is somewhat larger than most new U.S. housing, and it is located in a climate which has moderately high heating energy requirements.

For new U.S. housing as a whole, allowing for savings attainable in both heating and cooling energy requirements, I believe that a rough but reasonable ballpark estimate of easily attainable point-of-use energy savings, using simple and reasonably inexpensive methods, is 30 million Btu per year for each new house. Also, I believe that a rough but reasonable ballpark estimate of the ratio between point-of-origin and point-of-use energy

Why Not Just Build the House Right?

saving—for all heating and cooling of houses, averaged over the United States—may be around 3. That is, for every million Btu saved at point-of-use there is, on the average, a lessened withdrawal of around 3 million Btu at "point-of-origin." By point-of-origin I mean the collection of oil wells, gas wells, coal mines, and uranium mines that originally supply not only the point-of-use energy but also supply all additional energy associated with refining, transportation, and conversion losses of the point-of-use energy.

If the preceding two rough estimates are accepted as reasonable, then it appears that the potential for point-of-origin energy saving, in current new house construction, is around 100 million Btu per year for each new house. That is, on the average, each new house built now could have been built a little better, at reasonable extra cost, so as to require about 100 million Btu per year less energy from point-of-origin than it will require.

Suppose that during the next few years every new house were built a little better than they are being built today, so that each new house did realize that potential point-of-origin energy saving of roughly 100 million Btu yearly compared to current construction. From a national energy consumption standpoint, would the cumulative energy savings so produced really amount to much in relation to our total national energy consumption? After all, the yearly total amount of energy mined or imported into the United States is not measured in such "trivial" quantities as a hundred million Btu. Rather, it is measured in hundreds of millions of hundred million Btus. Confronted with such enormous figures, what is the big deal in building a new house a little better, even if it does save a hundred million Btu every year?

As a first step in getting a handle on the preceding question, a problem in elementary arithmetic, albeit with big numbers, can be posed:

If every new house is built a little better, so that is saves 100 million Btu each year at point-of-origin, if a new house lasts 30 years, and if 1.5 million new houses are built every year (which is about the number actually built yearly in the United States), then what will be the total energy saving, through their lifetime, of all the houses which will be built in the next 12 years?

The answer to that problem is $54 \times 10^{15}$ Btu, or 54 quadrillion Btu. That is, if every new house that will be built during the next 12 years were built somewhat better than they are being built today (instead of just the same), then the cumulative energy saving of that 12-year crop of houses, over their 30-year lifetimes, could be about 54 quadrillion Btu.

Raymond W. Bliss

One way to put that potential saving of 54 quadrillion Btu into meaningful context is to note that is is almost exactly equal to the total amount of oil energy presently believed recoverable from the giant North Slope oil field in Alaska. The total oil energy expected to be mined from that oil field, over its lifetime, is presently estimated at 56 quadrillion Btu (9.6 billion barrels of crude oil). Hence, in a very true sense if we can in some way manage to build all our new houses over the next 12 years somewhat better than we are now building them, we will have accomplished the same thing, energy-wise, as discovering and developing another oil field as large as the Alaskan North Slope.

Environment-wise, if we can in some way manage to build all those houses a little better, we will have accomplished much more than discovering and developing another source of energy as large as the Alaskan North Slope. The mining, transportation, and use of large amounts of energy, be it from oil, gas, coal, or atoms, beneficial as it is, also seems inherently to have aspects of strife and contention, and inherently to have aspects perhaps best described as creating a public nuisance on a global scale.

There is much to be said for—and quadrillions of Btu to be saved by— just building the house right in the first place.

## Notes

1. Charles V. Cuniff, "Solar Radiation on Walls Facing East and West," *Air Conditioning, Heating & Ventilating*, 55:10 (Oct. 1958), 82–88; and Cuniff, "Solar Radiation on Walls Facing North and South," *Air Conditioning, Heating & Ventilating*, 56: 8 (Aug. 1959).

2. U.S. Weather Bureau, *Local Climatological Data—Boston, Mass.* (Asheville, N. C.: The Bureau, 1965).

3. U.S. Department of Housing & Urban Development, *Minimum Property Standards—One and Two Family Dwellings*, HUD Pub. No. 4900.1 (Washington, D.C.: GPO, 1973, revisions of July 1974 and January 1975).

4. U.S. Department of Commerce, Bureau of Census, *Statistical Abstract of the United States* (Washington, D.C.: GPO, 1973), p. 684.

About 20 percent of the annual U.S. energy budget is used to provide space and water heating for buildings in this country by means of fossil fuel or electric systems. Low temperature solar thermal energy systems, providing direct heat at temperatures below 212° F, could be used instead of conventional systems, which would help to mitigate the U.S. energy dilemma. First, however, applications of low temperature solar thermal energy must become economically competitive and be widely used throughout the building industry in the United States. But there are obstacles and barriers which must be overcome before the use of solar energy in buildings is accepted, and therefore before solar energy can have a major impact on mitigating the current U.S. energy dilemma.

Recent analyses indicate that solar heating and cooling systems for residential buildings are nearly, but not quite, economically competitive with fossil fuel and electric systems.[1-5] But other related research indicates that even when solar space conditioning is economical institutional barriers to its widespread acceptance will remain.[6]

The main obstacle to the acceptance of a solar space conditioning system is the initial cost compared to that for a conventional gas, oil or electric system. However, although the solar system costs more initially, it has a lower operating cost than a conventional system. One way to compare solar with conventional space conditioning is to compute the payback period, that is, the number of years required for the solar system to pay for itself through its operating cost savings compared to a conventional system. A simplified example will illustrate this comparison.

A typical, conventional electric water heater for a house costs about $150 and lasts for about 10 years. The cost of electric water heating may be about $15 per month for a typical family of four. A solar water heating system costing about $720 could save two-thirds of the electricity cost and last for 15 years, the savings amounting to about $120 per year. Ignoring discount rate and maintenance for simplicity, the solar energy system has a 6-year payback period ($720 cost divided by $120 savings). Since the solar water heater will pay for itself well before it wears out, it would seem to be a reasonably economic investment. However, experience with a wide variety of products indicates that most consumers require a 5-year payback or less before they will purchase a product. Because at current production rates the high initial cost of a typical solar system will not be paid back in 5 years, an incentive is needed to facilitate the widespread use of solar space conditioning systems.

# 4 Barriers and Incentives to Encourage the Use of Solar Heating and Cooling
Alan S. Hirshberg

Other barriers reflect institutional characteristics of the building industry and its working environment: building codes constrain the types and uses of various building materials; the lack of sun rights makes a potential buyer think twice about installing solar energy if he thinks his neighbor might build a taller building to the south; the difficulties builders have in obtaining the extra construction capital needed for solar equipment; and adequate product performance information is not readily available.

In this essay I will examine such institutional barriers and other obstacles to the widespread acceptance and use of solar space conditioning systems, and explore some general policies which could help to overcome them. The potential impact of financial incentives will also be discussed; a tax incentive of 25 percent, for example, could speed the use of solar energy by 7 to 8 years and produce an 8 percent reduction in fossil fuel energy use by 1990. And, finally, I will present a preliminary—but what I

Alan S. Hirshberg

**Figure 4.1**
Solar collector for a new home in Maryland; expensive at first but cheaper in the long run. (Photo courtesy of the Aluminum Association, Inc.)

Solar Heating and Cooling

believe to be a reasonable—incentive package that could be helpful in promoting solar energy both at the federal and state level.

Even if the economic feasibility of solar heating and cooling technologies is proven, there are institutional barriers imposed on and operating within the building industry that could limit or slow the adoption of solar technologies. Because these institutional barriers have been described elsewhere,[6,7] only a brief overview of the more important issues will be presented here.

*Building Codes* Building codes imposed on the building industry can potentially constrain the adoption of solar energy. There are over 3,000 building code authorities in the United States, most of which have different codes. Unlike much of Western Europe, which has a federation of national systems of testing and evaluation laboratories (called "agrément systems") that reduce the effect of local code variations, technological innovations in the United States must often be separately approved by each code-setting body. This, of course, could limit the potential for cost cutting through mass production of standardized solar products.

In addition, most U.S. building codes are "specification" rather than "performance" oriented. Thus, they tend to constrain the use of new techniques since they specify the use of existing materials and practices. The successful builder is therefore usually reluctant to try a technological innovation that requires code modification. Given this serious constraint, to be successful the builder will probably require that the solar energy device reflect existing codes to the greatest extent feasible; for example, this may mean the exclusion of plastic pipe or other non-standard materials.

Building codes do not seem to pose a major problem for solar space conditioning at the present time. However, as Richard Schoen has pointed out, codes are often reactive and a few poorly designed or installed jobs could cause the creation of restrictive codes.

*Financing Constraints* Financial constraints on the building industry could also limit the adoption of solar technologies. Because it operates on borrowed capital, the industry is sensitive to fluctuations in interest rates and hence is a cyclical activity, depending delicately upon the supply and cost of capital. Depending upon the general scarcity of money and upon the total time of construction, financing charges can add 5 to 12 percent to the cost of new housing. Given narrow profit margins, these charges easily spell the difference between a builder's financial success or failure.

Alan S. Hirshberg

The industry is highly sensitive to initial investments (the first cost of its products), and the normal way to reduce the risk of high finance charges is to reduce initial capital requirements. Solar devices which have lower operating costs but higher initial investment costs than other energy systems could be expected to meet industry resistance. Also because the cost of capital is a large part of his operating expenses, a builder may tend to resist a solar product out of fear that he would lose money from the extended construction time needed for the new system. Another fear the builder may have is that he will not be able to pass on to the buyer the higher initial cost of a solar system, even if it has clear life-cycle cost advantages. Historically, buyers have been concerned primarily with initial or first costs. Coming up with a down payment for a house is difficult enough without having to pay an additional $1,000 to $2,000.

Another financing constraint is that lending institutions often narrowly restrict the choice of space conditioning equipment by setting cost limits based on the historical costs of conventional equipment. Since conventional equipment has lower first costs, such loan rules in effect exclude consideration of the solar option.

## Organizational Structure

The organizational structure of the industry produces institutional resistances to new technologies like solar energy. First, because of the variability of weather, building sites and codes, and the differences in individual tastes and life style throughout the country, the industry is regional. Regional differences require that flexibility of the design of solar devices should be "engineered-in" from the beginning. Second, the industry is highly fragmented. Of the industry's more than 300,000 builders, 90 percent produce less than 100 units per year. The largest builder produces less than 1 percent of the annual total. Furthermore, the industry is horizontally stratified, that is, it is comprised of many elements performing separate functions. No single person or organization is normally responsible for integrating all of the functions and controlling the residential construction process from beginning to end. Industry fragmentation and horizontal stratification combine to create broadly disaggregated markets and these, in turn, tend to slow the acceptance and diffusion of technological innovation.

In an environment exhibiting these organizational characteristics, those interested in the introduction and diffusion of solar energy devices are faced with difficult marketing, sales, and service problems. At a minimum, solar energy devices will have to achieve "product-fit" within the industry: new products must fit the existing industry distribution, sales, and service systems or, alternatively, be capable of establishing a parallel, equally effective system.

Two distinctly cultural aspects of institutional characteristics also shape the industry and must be accounted for in efforts to introduce changes. The industry is craft-based and operates through a series of individual craft unions that contribute separate skills and functions to the construction process. These unions have a great deal of control over acceptance of individual technological innovations. For this reason and because there is a relative absence of "performance specification," there tends instead to be a heavy reliance on previous "ways of doing things," and a general resistance to change. The result is a conservative social system which is also generally resistant to change.[7,8]

Factors such as building codes, financing arrangements, and the organizational structure of the industry pose constraints that tend to slow the pace of technological innovation in the building industry. Thus, general policies that will foster the acceptance and use of solar energy in the building industry must take into consideration not only the economic feasibility of solar technologies but also institutional factors such as the ones described here.

### Economic Competitiveness

At the present time, solar energy is nearly economically competitive with conventional fuels. For some applications, solar energy is competitive; for example, solar water heating is less expensive on a life-cycle basis than electric water heating in many parts of the country. It is not, however, currently competitive with natural gas for space heating. This is, in part, because of regulatory policies that insulate the consumer from the true marginal cost of acquiring new supplies of natural gas and force a low well head price on existing natural gas supplies.

Both gas price deregulation and adoption of a policy of marginal cost pricing by state and federal regulatory bodies would provide an incentive to conserve fossil fuels and make solar space conditioning more attractive to a buyer. The likely result of deregulation would be that the price of gas would be raised two- or threefold. (For example, intrastate natural gas in Texas, which is not subject to federal regulation, has sold recently for $1.75 to $2.25 per thousand cubic feet versus $0.70 to $1.00 per thou-

Alan S. Hirshberg

sand cubic feet in California.) Marginal cost pricing would require utility prices to be set at a level that reflects the cost of acquiring new supplies and would therefore make the user pay the full fair market price for his energy. Current utility pricing policy insulates the consumer from the true costs of new supplies by "rolling in" the price of the new supply with the price of existing supply contracts. An example should help clarify this point.

Some new natural gas supplies such as coal gasification are estimated to cost $4 per thousand cubic feet. Assuming $1 per thousand cubic feet distribution charges and a 50 percent efficient usage, the final cost of heat delivered to the consumer would be about $10 per million Btu. The cost of solar water heating delivered to the consumer is about $8 per million Btu, assuming an 8 percent loan and 20-year term. A "rational" consumer will choose to purchase the solar energy system since it will save him about $2 per million Btu. However, under current regulatory pricing practice the cost of the new gas supply will be "rolled in" with the existing supply that currently costs about $1.50 per million Btu. Assuming that the new gas supplies about 5 percent of the total (not an unreasonable amount for a coal gasification plant), the result of "rolled in" pricing will be that the gas will cost about $1.70 per thousand cubic feet, or $3.40 per million Btu delivered, assuming 50 percent efficiency. (If one assumes 100 million cubic feet at $1.50 and 5 million cubic feet at $5, the result is 105 million cubic feet at a total value of $175 million or about $1.70 per cubic foot.)

The rational consumer will find the $3.40 per million Btu "rolled in" natural gas a bargain, even though the cost of the new gas alone exceeds the cost of the solar space conditioning. Thus, the "rolled in" pricing method used by regulatory bodies essentially insulates the consumer from the true cost of new fuel supply. This pricing method is a substantial institutional barrier to the use of solar energy.

One alternative to marginal cost pricing for overcoming this barrier has formed a major premise of Project SAGE, which is funded principally by the Southern California Gas Company and the National Science Foundation and supported by the Federal Energy Administration (FEA) and the Energy Research and Development Administration (ERDA).

One aspect of Project SAGE is to examine the notion of utility ownership and maintenance of solar energy equipment. Under this arrangement, a utility will own the solar equipment, and the equipment would be rolled into the rate base. This is equivalent to treating solar energy on an equal footing with new natural gas and electric supplies. The consumer would be charged on his monthly utility bill just as he is now for conventional energy use. Under a set of rules, proposed by E. S. Davis, the utility would not be allowed to manufacture or install these systems itself, but would rely on the normal building industry channels and the solar equipment manufacturers and installers for these services. Utility ownership would overcome the "rolled in" pricing barrier and would substantially reduce the high initial cost problem.

Even if solar energy is competitive, it may not be used in buildings because of the barriers to life-cycle costing described above. But regulations could be adopted to require each builder or building owner to examine the life-cycle cost of the energy systems for the building and to use the system with the lowest life-cycle cost. In this way the solar energy system, which has a higher initial cost but a lower operating cost, would be comparable and on an equal footing with a conventional heating, ventilating, and air conditioning system, which usually has a lower initial cost but a higher operating cost. Such a policy has been developed by the California Coastal Zone Conservation Commission, which requires all new projects to submit a study indicating the feasibility of solar energy and its life-cycle comparison with conventional space conditioning systems.

This policy has the advantage of not requiring the use of solar energy in projects where solar energy is not feasible because of obstructions or other problems, or where solar energy life-cycle costs are not competitive with conventional sources. Unfortunately, this type of policy has encountered stiff resistance from developers who either do not understand life-cycle costing or who fear that they would not be able to sell the life-cycle advantage to their customers. In addition, this policy is difficult to administer, in part, because the life-cycle cost analysis depends upon assumptions concerning fossil fuel price escalation, solar energy maintenance schedules, and discount rates, etc. Each of these assumptions contains enough uncertainties to change the calculated life-cycle cost of solar energy by a large amount and is subject to manipulation.

Both the deregulation and life-cycle costing policies have encountered political opposition. Consumers resist further price increases and builders wish to avoid the administrative and other perceived problems with life-cycle costing. In this political environment direct cost reducing policy should also be considered. Both tax incentives and low interest loans can be effective in promoting the use of new technologies like solar energy and overcoming such resistance.

Alan S. Hirshberg

*Tax Incentives* Tax incentives include tax abatement, tax credits, and accelerated depreciation allowances. Tax abatements include the exemption of solar energy equipment from property tax and/or sales tax. Over 100 pieces of legislation to promote solar energy have been introduced in 32 states. The net effect of property tax abatements is the reduction in annualized cost of solar systems equivalent to about 7 to 10 percent. Although the annual financial impact is small, the political value of having such abatement legislation is high because it puts state legislators on record as saying that the use of solar energy is in the public interest.

Tax credits usually provide a reduction of income tax up to some percentage of the total first cost of the solar equipment with a maximum tax credit ceiling. Business firms may choose an accelerated depreciation schedule rather than a direct tax credit. Several tax credit bills have been introduced into Congress, giving tax credits from 10 to 50 percent. The Energy Conservation and Conversion Act of 1975 (HR 6860), which passed the House of Representatives in the spring of 1975, would allow a homeowner to reduce his income tax by as much as $2,000 or 25 percent of the cost of any solar energy system costing up to $8,000. For commercial builders, a 10 percent investment tax credit on solar equipment would be allowed. Or, if a commercial builder is more concerned with life-cycle costs instead of capital outlays, he may choose instead to depreciate the value of the solar system at the rate of 20 percent per year. This 5-year depreciation will provide a profitable alternative to the investment tax credit when the useful life of the systems is 12 years or more. Both incentives can be used until January 1, 1981, when they will expire. The Senate version of HR 6860 was stalled in the Senate Finance Committee. However, a similar bill, HR 10612, passed the House in the spring of 1976 and it is expected that it will be deliberated in the Senate this fall.

*Low Interest Loans* Low interest loans to provide financing for solar space conditioning at rates which are below commercial rates can be an effective incentive. In general a low interest loan program should require at least a small down payment so that the potential buyer would have a financial stake initially in the solar system. Exceptions perhaps could be allowed for low income buyers.

Low interest loan legislation has been introduced in the Congress, and several states are also considering using their bonding power to raise money for state financed low interest loans. The interest rates on these loans vary from as low as 2 percent to about 6.5 percent.

The tax credit is a more direct way of providing an incentive than a low interest loan and should have a lower administrative cost. However, the tax credit could require a large amount of funding initially and for this reason may meet political resistance. A low interest loan will not have this initial funding requirement. If the loan is administered through the existing financing channels, administrative costs and red tape would conceivably be minimized. Of course, these direct financial incentives must be administered cautiously to avoid or minimize the potential negative effects of such policies.

In the absence of incentives, solar space conditioning systems would require a long time to be developed to a point where they have a substantial economic advantage compared to conventional fossil fuel or electric systems. But while incentives would tend to encourage the early use of solar energy they would also tend to discourage entrepreneurs from searching for better or cheaper techniques for applying solar energy technologies. The incentives might also actually slow down the use of solar devices by making people who might otherwise buy a solar energy device wait for enactment of some kind of incentive. Potential solar users might reason that Congress is just about to provide a tax incentive for the use of solar energy and will wait until the incentive is enacted before buying. If the political process delays the enactment of the tax incentive, the proposed tax incentive will, in effect, delay the use of solar energy. Finally, it is important that the incentives be phased out as the solar industry matures. Ideally, all policies should be formulated in such a way that the incentives would automatically expire when the industry is well established.

Policies that promote the gathering and dissemination of information, both passive and active, would serve to reduce the burden on developers and other potential users in learning about the new solar technologies. Passive methods include the development of data banks and clearing houses with information concerning solar energy. A clearing house, for example, could provide a central place where developers could locate names of manufacturers and designer/installers for solar energy equipment.

By keeping one central source, the costs of finding out about solar energy would be reduced. Both ERDA and FEA, as well as the Solar Energy Industries Association (SEIA) and a number of other groups, maintain a list of solar energy firms. In addition, the U.S. Department of Housing and Urban Development is required by Public Law 93–409 (the Solar Heating and Cooling Demonstration Act of 1975) to maintain a solar heating and

Alan S. Hirshberg

cooling data bank. Unfortunately, the clearing house notion usually does not go far enough in reducing the informational cost to the prospective users. In order to utilize it, users must know about the clearing house, must know how to ask the right questions, and must then be able to sort through and understand the data provided them. For a solar energy novice, this can be quite time consuming and confusing.

For this reason active information dissemination methods are likely to produce superior results. These would include focused dissemination activities and would require an intermediary between the solar energy community and the user community. The intermediary would help translate solar energy technical information into a form that is understandable to builders and other potential users. The intermediary could track and report problems concerning the users' application of solar energy back to the solar community. The intermediary would utilize the appropriate media for each group (for example, *Professional Builders, AIA Journal,* or the *ASHRAE Journal* of the building industry as well as the *Journal of Solar Energy* or the SEIA Newsletter of the solar energy community). Fortunately, the responsible federal agencies seem to be aware of this problem and have attempted to design active dissemination programs.

## Risk Reducing Policies

Another class of policies could help reduce the risk of being one of the initial users of solar energy. Prior to 1974, over 100 buildings have been built in the United States which utilize solar energy, but most builders probably believe that the use of solar energy systems poses substantial technical risks. Whether these perceptions are factual or not is not the primary issue; it is sufficient to know that this is what builders and other potential users believe.

The perception of risk could be reduced if potential users were convinced of the success of initial projects. An independent project evaluation and policy analysis capability, combined with the type of active information dissemination program described above (a concept characterized as an Implementation Center), can provide a way of reducing the risk of solar energy to builders. A similar concept, called the Energy Extension Service, has been developed by ERDA, and legislation (S 3105) proposing the creation of such a service has been proposed by Rep. Ray Thornton of Arkansas. The Service, like the Implementation Center, would operate along the lines of the Agricultural Cooperative Extension Service and provide relevant information to potential users of solar energy and energy conservation technologies.

Demonstration projects such as those provided in the Solar Heating and Cooling Demonstration Act could also provide the necessary information, on a regional basis, to show builders and others that the risks in using solar energy systems in their buildings will be minimal. Moreover, demonstration projects could help to create markets for solar components and thus would be an incentive for manufacturers.

Policies that improve economic competitiveness, promote life-cycle costing, provide direct financial incentives, facilitate the dissemination of information, and reduce the perceived and/or real risk in using solar space conditioning are all useful in stimulating early acceptance and widespread use of solar space conditioning systems. Careful orchestration of these policies will help to provide the climate necessary for smoother acceptance of solar space conditioning and to remove many of the principal institutional barriers to its use.

What many people want to know about solar space conditioning is how much of an impact we can expect, and how soon can we expect it? To answer these questions it is useful to suppose hypothetically that policies are adopted that are adequate to remove the institutional barriers, such as the ones described here, so that the rate of acceptance and the use of solar space conditioning can be determined primarily on economic grounds.

In order to better understand the rate of adoption, a market penetration model was developed at the Jet Propulsion Laboratory for Project BASE, sponsored by the Southern California Edison Company.[9] The model provides for competition between seven solar space conditioning systems with energy conservation and conventional energy systems with energy conservation. Energy conservation was included in both conventional and solar systems because it makes the most economic sense and avoids the "breakfast of champions effect": that is, a case in which 80 percent of the benefits came from the first 20 percent of the additional cost.

The model was prepared for a variety of residential and commercial buildings. Over 50 solar space conditioning systems were designed and costed using the National Construction Estimator.[10] The model provides for no market penetration until a solar energy system achieved a 5-year payback or an 18 percent internal rate of return. Once this criterion was achieved, penetration was allowed at rates consistent with the historical rate of penetration of new technology in the building industry.[11] The mo-

Alan S. Hirshberg

del was able to help answer two important questions: (1) What amount of financial incentive should be given to solar space conditioning? (1) What will be the impact of the financial incentives?

## Assessing the Impact

In order to assess the impact, it is convenient to determine the impact of financial incentives upon the first cost of the solar systems. For example, property tax abatement reduces the effective first-cost by about 10 percent. (This is based on a 25 percent tax assessment and a $10 per $100 tax rate that is probably typical of most cities. The effective first-cost reduction is the ratio of the annual saving of the property tax abatement (about two percent) to the capital recovery factor that is approximately 0.2 for a 15 percent loan for 10 years.)

Tax credits produce first-cost reductions equal to the percentage of the credit, and low interest loans reduce the first cost as the capital recovery factor is reduced. (The capital recovery factor, $r(1 + r)^n/([1 + r]^n - 1)$, where $r$ is the relevant interest rate and $n$ is the period of the loan, specifies what portion of the loan and interest must be paid each period.) Thus, for example, with a 20-year loan period, the first cost would be reduced more than 30 percent if the interest rate were reduced from 10 to 5 percent.

The effect of direct financial incentives can be assessed in two ways. First, their impact can be assessed in terms of speeding the development of a solar energy industry. Once the industry has been developed, the price of solar collectors will probably decline substantially, though perhaps not as dramatically as other new products. This price decline, however depends, in part, upon the growth of orders for collectors. Here, we are confronted with a chicken and egg problem.

Prices of collectors must decline before many square feet of collectors will be competitive enough to be sold. A major way of reducing prices is to stimulate large orders so that manufacturing economies of scale can occur. A rational way to establish an incentive level would be to choose that incentive level which reduces the cost of installed solar systems to the price that would be in effect if the industry were fully established.

Studies by E. S. Davis[9] indicate that a collector price of $3.00 per square foot (in 1974 dollars), FOB the factor, is reasonable for an all glass collector with selective absorber, given several million square feet per year of sales. If the cost of collector installation and reasonable overhead and profit rates are added to this, the estimated equilibrium price of collectors could be about $5.00 per square foot installed. With minor exceptions (such as controllers and sensors), this times the total square feet of installed collector area would be the cost of the nonconventional part of an installed solar energy system.

The other costing of the system would be for piping, storage tanks, installation, insulation, and other conventional components for which prices can be well established using manufacturers' price lists and the National Construction Indicator. (The percentage of conventional to collector cost ranges from 50 to 25 percent depending on the size of the system. An empirical equation for the total cost $(C)$ of the system as a function of the collector area $(A)$ has been determined to be: $C = pA + dA^b$ where $p$ is the price per foot squared of the collector installed and $d$ and $b$ are constants. The details of the solar collector estimation method is described in reference 9.)

### A Solar Energy Incentive Package

Incentives at both the federal and state level are needed if we want to gain the acceptance of solar space heating and cooling systems in the building industry in the United States, given the organizational barriers in the industry. And, even when the operating costs of a solar space conditioning system become less than those for the conventional systems now being used in this country, a main obstacle to the acceptance of a solar system is its initial cost compared to that for a conventional gas, oil, or electric system. An incentive is needed to overcome this economic barrier.

A federal incentive package designed to improve the economics of solar space conditioning could include two primary incentives: tax credits (or equivalently accelerated depreciation allowances), and low interest loans. A 5 percent loan for 20 years would yield a 25 percent equivalent first-cost reduction compared to a typical 9 percent mortgage loan for the same 20-year term. Depending on the alternative fuel and location, the monthly payments on the loan plus the monthly cost of the additional fuel could be less than for the monthly fuel costs without the solar system. Establishing a low interest loan rate tied to federal money could help make the level of incentive remain relatively constant. In addition, a direct tax credit of 25 percent could be given initially to provide additional inducement for solar energy.

The use of this combination of incentives would provide a 50 percent

Alan S. Hirshberg

incentive, the level required to make the total cost of typical space conditioning systems equal to the expected costs after the solar industry matures. After 1980, the tax credit could be terminated which would scale the total incentive from 50 percent to 25 percent. After 1985, the low interest rate could be gradually raised until the entire incentive were phased out by 1990.

Two additional policy actions to improve the economics of solar space conditioning and to reduce the first-cost barrier are necessary at the state level. States could provide a property tax abatement that would remove the added value of solar space conditioning from the assessed value of the property. This would provide about a 10 percent reduction in the annualized cost of solar system and would demonstrate tangible state support for solar energy. In addition, state regulatory bodies could help establish the option of utility ownership of solar space conditioning equipment. Allowing consumers the choice between buying the solar space conditioning equipment themselves—utilizing the proposed federal incentive package—or obtaining the equipment from a utility may help keep solar costs low by promoting competition between installers who are contracted by the utility and installers who are not. Utility ownership will help mitigate the first-cost barrier.

Finally, both federal and state governments can expand the demonstration program and establish Implementation Centers. Feasibility studies for demonstration projects should include a life-cycle cost analysis, including a cost reduction at the level of the incentive prevailing at the time of analysis.

Incentives, such as the ones presented here, could stimulate savings in fossil fuel use and boost the impact of solar energy on the U.S. economy.

## Incentive Level

A series of curves is presented in figure 4.2 that equate the incentive necessary to reduce the installed solar system cost, using currently available collectors at their current prices, to the system cost if the collectors were priced at $5 installed (in 1974 dollars). The curves are drawn for typical residential water heating and space heating solar systems, ranging in size from 100 square feet to 1,500 square feet.

An example should help clarify the meaning of figure 4.2. The 1974 price of installed solar collectors (not including storage, controls, etc.) is roughly $15 per square foot. If the size of the system is 500 square feet,

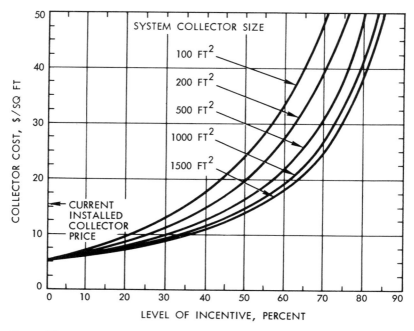

**Figure 4.2**
Solar collector cost with incentive. This series of curves equates the incentive necessary to reduce the cost of an installed solar system for space and water heating, using currently available collectors priced in dollars per square foot, to the system cost if the collectors were priced at $5 installed.

then the amount of incentive that will make the total system cost equivalent to the expected long term cost in a mature solar industry is about 45 percent.

Considering a range of sizes, we observe that incentives between 35 and 55 percent on the total installed systems with the 1974 collector technology, and production rates will produce total installed systems costs that are the same as for the eventual $5.00 collector after the solar industry has matured. This provides a reasonable level of incentive for solar systems based on the need to spur the growth of a large enough industry to produce cost reductions based on economies of scale in manufacturing.

A second way of establishing reasonable incentive levels is to examine the energy saving produced by alternative incentive levels and to choose the level which meets a preestablished energy savings goal. The energy sav-

Alan S. Hirshberg

ing produced by three different first-cost reducing incentive levels (0, 25, and 50 percent) as a function of time shown in figure 4.3. The curves were developed to estimate the market penetration of solar energy systems in single family units in southern California under the assumptions that natural gas prices remain fixed in constant 1974 dollars, while electric prices rise at 4 percent per year above inflation from a 3.5 cents per kilowatt hour level in 1974. An embargo on new natural gas hook-ups after 1978 is also assumed.

From figure 4.3, one can estimate the difference between no incentive (labeled A), a 25 percent incentive (labeled B), and a 50 percent incentive (labeled C). With a 50 percent incentive, a 10 percent energy saving is achieved by 1985 while the no incentive case produces only a negligible saving. Furthermore, a 25 percent incentive "buys" us the same energy displacement as the no incentive case but does this 7 to 8 years earlier, while the 50 percent incentive case "buys" about 6 to 7 years earlier penetration than the 25 percent case.

Although these estimates were developed for the single family market in southern California, they may roughly reflect the market penetration potential in other parts of the country for several reasons. The economics of solar space and water heating are more favorable for the multiple family market. New multiple building has averaged nearly two-thirds of all new residential units in the last five years and the trend is likely to continue. Natural gas hook-up embargoes exist in several parts of the country, and the price of natural gas and electricity in some areas is higher than the $0.10 per therm, $0.035 kilowatt hour, rates paid in southern California. Furthermore, the analysis assumed that the single family purchaser would require a minimum 5-year payback period. Analysis of the adoption of water heating in Florida by Scott and others[12] indicates that a 5-year payback was sufficient for purchasers in the 1940s. The institutional market, including schools and hospitals, may be expected to find even longer payback periods acceptable.

Many types of incentives are available, each of which could be helpful in promoting solar energy. The specification of a package which combines these incentives is a useful task. I have developed what I consider to be a reasonable incentive package. Although the package is preliminary and therefore not analytically precise, it meets the requirement of policy science[13-15] by providing a framework for discussion during the time when policy-makers are debating related issues and need even first-cut estimates.

Many different incentive packages could be developed. Different mea-

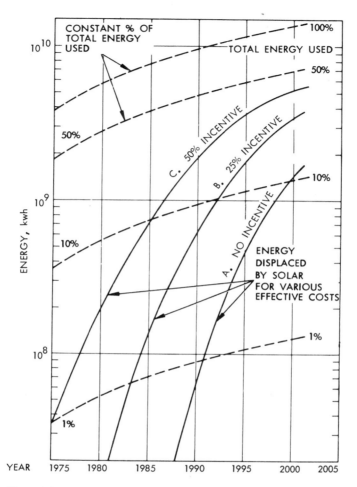

**Figure 4.3**
Energy savings with solar space and water heating. The energy savings produced by various incentive levels to reduce the first cost of solar space and water heating is shown here. Though these curves were developed to estimate the market penetration of solar energy systems in single family units in southern California, using a Jet Propulsion Laboratory program, they may roughly reflect the market penetration potential in other parts of the country. The dashed curves show fractions of the total energy used in the southern California market indicating the degrees of energy savings that might result from the different incentive levels.

Alan S. Hirshberg

sures are probably needed for a coherent policy, and guidelines will need to be established for phasing the various elements of the policy in and out over time. The extent to which direct financial incentives are offered will depend on how society values the fuel savings possible with a solar program. The package I have proposed here, which could yield a 10 percent annual fossil fuel savings by 1985, is illustrative of what I consider to be a politically feasible and effective way of encouraging early and widespread use of solar energy.

## Notes

1. George O. G. Löf and R. A. Tybout, "Cost of House Heating with Solar Energy," *Solar Energy*, vol. 14, 1973.

2. E. S. Davis, "Project SAGE Phase O Report," Environmental Quality Lab Report No. 11 (Pasadena, Calif.: California Institute of Technology, 1974).

3. General Electric, *Solar Heating and Cooling of Buildings*, a report to the National Science Foundation, June 1974.

4. TRW, Inc., *Phase O Report on Solar Heating and Cooling of Buildings*, June 1974.

5. Westinghouse Corp., *Solar Heating and Cooling of Buildings*, June 1974.

6. Alan S. Hirshberg and Richard Schoen, "Barriers to Widespread Utilization of Residential Solar Energy: The Prospects for Solar Energy in U.S. Housing Industry," *Policy Sciences*, Dec. 1974.

7. R. Schoen, A. S. Hirshberg, and J. Weingart, *New Energy Technology for Buildings* (Cambridge, Mass.: Ballinger, 1975).

8. D. Schon, *Technology and Change* (New York: Delacourte Press, 1967).

9. E. S. Davis, R. A. French, and A. S. Hirshberg, "Scenarios for Utilization of Solar Energy in Southern California Buildings," report 5040-10 (Pasadena, Ca.: Jet Propulsion Laboratory, California Institute of Technology, Aug. 1975).

10. E. S. Davis and L. C. Wen, "Solar Heating and Cooling Systems for Buildings: Technology and Selected Case Studies," JPL report 5040-9, July 1975.

11. A. S. Hirshberg, "Representative Buildings for Solar Energy Performance Analysis and Market Penetration," JPL report 5040-3, Sept. 1975.

12. J. A. Scott, R. Melicher and D. Sciglimpaglia, *Demand Analysis Solar Heating and Cooling of Buildings:* Phase 1 Report to National Science Foundation (Washington, D.C.: GPO, 1974).

13. Y. Dror, *Design for Policy Sciences* (New York: Elsevier, 1971).

14. A. Etzioni, "Policy Research," *American Sociologist*, 6 (1971), 8–12.

15. H. Lasswell, "Policy Sciences," *Policy Science*, 1:1 (Nov. 1972).

# III  Solar Electricity

One of the dark horses in America's stable of technical fixes to the energy crisis is to generate electricity by the direct use of solar energy in large central power plants. As with all forms of solar energy, the basic difficulty in achieving economic viability is the diffuse nature of the energy source. As the most intense solar insolation is only 1 kilowatt per square meter, it is necessary to use large collection areas to capture the energy. And it is this factor that gives rise to the most obvious characteristic of solar power systems—they are large and capital intensive. Solar energy is also intermittent both on a regular (diurnal) and irregular basis (weather), at least on the earth's surface. This requires the solar power plant to be even larger: it must over-collect energy during the day and store it for use during the evening. Energy storage is a relatively undeveloped area, and only lead-acid batteries and pumped hydro are widely available technologies at this time.[1]

The decision to use or not to use solar-thermal-electric power plants on a large scale in this country will be made on the basis of its total social acceptability to our society in relation to the alternatives. Moreover, for widespread acceptance the net benefits for this particular solar option— that is, the benefits less the combination of direct "consumer" costs and the perceived social costs due to anticipated health effects, federal research costs, environmental impacts, and other social costs—must be greater than the net benefits perceived for nuclear- or fossil-fueled power plants. Thus, if we are to use solar thermal energy as a national energy source, we must ask ourselves some of the following questions:

- What is the economic cost of a solar-thermal-electric central power plant?
- How will it compare with the cost for a nuclear or a coal plant toward the end of the century, when the initial series of commercial solar plants may be built?
- Where is this energy source located?
- How large a resource is it?
- Can it be used throughout the United States?
- And, finally, what are the social costs of using this energy source?

Both "good" sun (about 5 kilowatt hours per square meter per day) and low cost-low use land are available primarily in the southwestern region of the United States. This region which includes eight states with a total land area of one million square miles is one-third of the continental United States. These considerations suggest that the production of electricity in large central power plants by the direct use of solar energy may only have a regional impact in ameliorating our energy difficulties.

# 5 Solar Power Plants: Dark Horse in the Energy Stable
Richard S. Caputo

To overcome this apparent regional constraint there must be, first, enough sun to generate enough electricity for this energy resource to have a national impact and, second, the energy must be transportable outside of the Southwest. There certainly is enough land to meet our needs on a national scale. Of the one million square miles of land in the sun bowl, about 2 to 16 percent is potentially available and suitable for use as a solar power plant.[2] This is 4 to 32 times larger than the area needed to generate the current national electricity requirements. Also the use of high voltage overhead direct current electric transmission on exclusive right-of-way can move energy thousands of miles at low cost (about 5 mills per kilowatt electric hour per 1,000 miles of transport) and high efficiency (about 95 percent).[3]

## Cooling Techniques

But while there would likely be plenty of land for a large-scale solar thermal-electric development in the Southwest, the scarcity of water is a constraint that cannot be overcome with conventional power plant cooling technology. For all practical purposes, there is no water available in the Southwest region for power plant cooling. The only rivers are the Colorado and the Rio Grande, which are overcommitted now. Wells are the only other source of cooling water indigenous to the region, but these will not support a sufficient number of solar-electric power plants to meet our national power needs, at least using current cooling techniques. These limited resources could be conserved, however, if dry cooling towers were utilized, though a capital cost and operating efficiency penalty of about 10 percent would be incurred.

The possibility of substantial development of solar thermal-electric generation outside the Southwest should not be dismissed prematurely. The "local" sun in other parts of the United States may almost be as good as in the Southwest.

On the middle Atlantic seaboard, for example, the total normal solar energy is only two-thirds of a good Southwest location, yet the relative performance is about 80 percent as good[4] as a place such as Inyokern, Ca., in the Mojave Desert.

This would increase the energy cost by about 25 percent and provide an upper limit to the acceptable costs for a long distance transmission link. And it may be feasible to overcome this cost penalty by deploying solar conversion systems at or near the distributed points of end use. This would

Richard S. Caputo

eliminate transmission costs and open the possibility of further cost cutting by using normally wasted heat to meet space conditioning, hot water, and process heating needs.

Such possibilities should be carefully evaluated. The discussion here, however, will be limited to large central solar-thermal-electric power plants with dry cooling towers in the arid Southwest that use long distance transmission via overheat direct current lines.

The federal government (DOE) is vigorously pursuing this potential national energy resource. By the year 2020, DOE estimates that 2.5 percent of total U.S. energy use, or 10 percent of the energy provided by alternative energy sources, could be provided by this type of a solar power plant.[5] The funding for solar-thermal-electric energy went from $1 million in fiscal year 1972 to about $20 million or nearly 10 percent of the total solar research and development budget in fiscal 1976. The fiscal year 1978 solar thermal budget is now at $60 million, or 11 percent of the total solar research and development, including biomass, hydro, and geothermal. In turn, this is 14 percent of the $3.5 billion Department of Energy budget, not including $0.5 billion for uranium enrichment. Thus, this solar thermal research is now 1.6 percent of federal energy research and development. Federally supported research and development for this technology has more than doubled every year since fiscal 1971.

Current planning shows a 10 megawatt (electric) plant coming on-line at Barstow, Ca., in 1980, with the first 100 megawatt (electric) commercial-size demonstration plant producing power in 1985.[6] This would have the same significance as the Shippingport, Pa., nuclear power plant of the mid-1950s.

## Power Tower

The specific approach being supported is the central receiver (power tower) concept where thermal energy is optically reflected from an array of mirrors to a central receiver at the top of a rather tall tower (100 to 600) meters). This receiver takes the place of a fossil boiler or a reactor core in conventional plants and can generate superheated steam. A fairly traditional steam power plant can be used to convert the steam into electric power in a conventional Rankine steam cycle.[7] This is the power plant equivalent of Archimedes fabled tale of using soldiers' shiney shields to reflect solar energy to burn the black sails of the invading Persian fleet in

Solar Power Plants

500 B.C. An artist's rendition of the central receiver is shown in figure 5.1 and illustrates the optical collection of energy to a central point.

The decision to concentrate the solar-thermal-electric program on a central receiver type of solar thermal power plant was made in late 1974 on the basis of preliminary economic assessments of alternative systems performed for the National Science Foundation. The costs estimated for alternatives ranged from 20 percent to 100 percent more than the costs estimated for the central receiver system. The unique feature that differentiates the central receiver system from the other solar power technologies is the use of optical means to transport the solar energy collected over a large field of collectors to a central point.

Alternative concepts involve the collection of solar energy throughout the collector field, that is, in distributed receivers. Two major choices exist: the heat can be moved to a central energy conversion plant via a fluid or disassociated chemicals pumped through a piping network[8]; or the heat collected can be converted to electricity in a small heat engine-generator and the electricity carried to a central point via wires. Thus, the distributed receivers can have either distributed energy conversion (small heat engines) or central energy conversion (large heat engine).

Several types of distributed collectors were considered in the initial NSF assessments: nontracking flat plate collectors, single axis tracking concentrators, and two axis tracking parabolic dish collectors (see figs. 5.2–5.4). The most attractive for a central electric power plant was found to be the parabolic dish (see figure 5.4). For simplicity, this type of distributed receiver is referred to as the "power dish." In the distributed receiver concept, solar energy can be collected throughout the collector field and transported by steam or chemicals to a central power plant where it is converted to electricity; or electricity can be generated, using a small heat engine, at or near the collector site and then directly transported as electricity. In the central receiver concept (illustrated in fig. 5.1) the thermal energy is optically reflected from an array of mirrors to a "central receiver" at the top of a rather tall tower (the "power tower"), and the energy is converted to electricity at a large central power plant.

### Solar Plant Costs

*Central Receiver* The results of recent detailed cost analyses for the central receiver power plant are shown in table 5.1. The capital cost breakdown is

**Figure 5.1**
A field of heliostats reflect the sun's rays to a central receiving tower in this artist's conception of a Boeing solar collector system. Heat in the tower would be used to produce electrical energy.

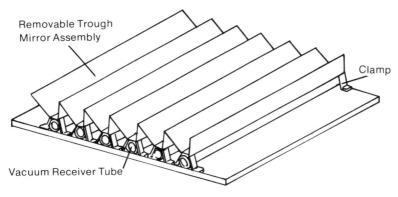

Removable Trough
Mirror Assembly

Clamp

Vacuum Receiver Tube

**Figure 5.2**
Non-tracking vee-trough "flat plate" collector. Energy can be transported by saturated steam, and converted to electricity at a large central steam or organic Rankine power plant.

Solar Power Plants

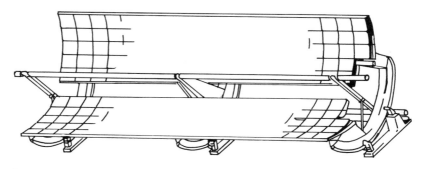

**Figure 5.3**
Single axis tracking parabolic trough collector. Energy can be transported by super-heated steam, and converted to electricity at a large central steam power plant.

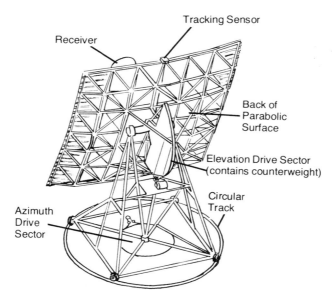

Tracking Sensor

Receiver

Back of
Parabolic
Surface

Elevation Drive Sector
(contains counterweight)

Circular
Track

Azimuth
Drive
Sector

**Figure 5.4**
Two axis tracking parabolic dish collector. Energy can be transported by steam or chemicals and converted to electricity at a central power plant; or electricity can be generated, using a small heat engine, at or near the collector site and directly transported as electricity.

Richard S. Caputo

given for a solar plant with an annual load factor of 0.55 including a plant availability of 0.86. The plant uses dry cooling to minimize water use. Cost estimates are projected to a mass produced, commercial basis assuming this future plant could be built in 1975. 1975 dollars are used. The greatest uncertainty exists for the heliostat direct costs, and a range is shown from $62/m² (near the DOE goal) to $145/m² (recent Jet Propulsion Laboratory estimate).[9]

The total construction costs vary from $1,500 to $2,200 per kilowatt (electric) for solar power plants with 6 hours of storage capacity at 70 percent of rated power. The dry cooling towers will be required by the year 2000 in the arid Southwest. The total heliostat surface is just over 1 square kilometer for a 100 megawatt (electric) plant with storage. Thus, the total plant cost is estimated to be about $200 per square meter of mirror surface. To put this number in perspective it is comparable to the current cost of housebuilding.

*Distributed Receivers* Similar studies have been done for the other solar power plant options. Their total construction costs compared to those of the central receiver are shown below:

*Power Tower:* Central receiver                     $2,260/kWe

*Power Dish:* Distributed receiver
    Steam transport                     $2,700/kWe
    Chemical transport                 $2,400/kWe
    Stirling engine with electric transport        $1,800/kWe

These latest analyses indicate that some of the power dish systems may be about as costly as the power tower for a central power plant.

The parabolic dish-chemical transport approach has a built-in storage capacity in the unreacted gaseous chemicals that could be stored very inexpensively if pumped underground.[10] Also the size of the solar plant could be increased from about 100 megawatts (electric), which is optimum for the central receiver, to 500 megawatts (electric) or more. This would introduce cost savings due to effects of scale. Control and central plant operation would be easier, and there would be more standardization in plant design.

The other power dish approach of interest is the parabolic dish-small heat engine concept. The power plant is essentially made up of many 10 to 25 kilowatt (electric) power plants. A small, light, inexpensive and efficient Stirling engine at the focal point of a parabolic dish collector may even be

Table 5.1

Central receiver solar power plant, 100 megawatts (electric), 6 hours storage capacity at 70% of rated power (70MWe), annual load factor = 0.55,[a] 1975$, 1975 plant start-up, dry-cooling, land area = 3.42 km$^2$ (1.32 mi$^2$)

| | Low | Mid | High |
|---|---|---|---|
| Direct cost | | | |
| Heliostat, $/m$^2$ | 62 | 91.4 | 145 |
| Receiver, transport, $/kWe | — | 252 | — |
| Energy storage, $/kWeh | — | 60 | — |
| Energy conversion, $/kWe | — | 250 | — |
| Total capital cost,[b] $/kWe | 1,510 | 1,825 | 2,260 |
| Operation and maintenance, 10$^6$ $/yr | — | 2.9 | — |
| Energy cost,[c] mills/kWeh | 70 | 85 | 105 |

[a] Including plant availability of 0.86.

[b] Includes direct costs, interest during construction, spares and contingency.

[c] The energy cost is the levelized (average) cost over the 30-year expected plant lifetime with an 8% cost of money. See a description of the economic methodology in J. W. Doane, et al., "The Cost of Energy from Utility Owned Solar Electric Systems," JPL–5040–29, Jet Propulsion Lab, June 1976.

self-starting when tracking the sun.[11] It would be more easily air cooled without cooling water consumption and be able to use irregular sized land.

When produced in an automotive-type engine plant at approximately 300,000 units per year, the Stirling engine cost should approach $1,000 for 25 kilowatt (electric) capacity. This is only $40 per kilowatt (electric) for the basic energy conversion component. A small Brayton engine mounted at the focal point would have similar performance and advantages as the Stirling engine, but the plant energy cost would be about 10 percent more expensive.[12]

Solar power plants based on the dish-small heat engine approach can also be scaled up to greater than 500 megawatts (electric); but perhaps more importantly, they can be scaled down literally to 20 kilowatts (electric). These small engines could be used not only for central power plants in the Southwest desert, but also for community-level power plants providing perhaps heat as well as electricity in total energy systems.[13] Also with such a modular design the growing demand for power could be met by gradually expanding plant capacity as the demand develops. The problems of tying up construction capital and making a commitment for a large plant 10 years before it will be needed would therefore be minimized.

Richard S. Caputo

**Figure 5.5a**
Solar reflectors at the Sandia Laboratories test facility in New Mexico. Each reflector in the bank is 9 × 12 feet. (Photo courtesy of Sandia Laboratory.)

In light of such potential advantages of the power dishes it would be of interest nationally to develop such systems as a back-up or even as an alternative to the central receiver (power tower) concept. However, the lion's share of DOE's central plant funds is allocated to the power tower concept.

### Conventional Costs

As shown in table 5.1 the cost of producing electricity today with a solar central receiver system would be 70–105 mills per kilowatt hours. To what should these costs be compared? The cost of today's utility power varies from approximately 20 to 100 mills per kilowatt hour (electric) as the plant load factor goes from baseload to peaking. Rather than considering today's cost, it would probably be of greater value to look at likely advanced conventional plants that will become available toward the end of this century. This is the time frame where solar plants may become in-

Solar Power Plants

**Figure 5.5b**
A heliostat, consisting of 25 four-foot square mirrors, used to focus the sun's rays at the Sandia Laboratories test facility in New Mexico. (Photo courtesy of Sandia Laboratory.)

Richard S. Caputo

creasingly available assuming the commercial-size demonstration plant comes on-line in 1985.

A typical nuclear plant available toward the end of the century will be a light water reactor very similar to today's plants. The first few liquid metal fast breeder reactors may also be available by the end of the century, but the unit costs may not be very different than those for the light water reactor. Possible advanced coal-based plants would include:

● a conventional coal-fired steam cycle plant with a limestone scrubber, which could remove after combustion about 90 percent of the sulfur;
● a plant with fluidized bed combustion and a combined-cycle of steam and gas turbines, where about 95 percent of the sulfur could be removed during combustion; and
● a combined cycle plant used in conjunction with low-Btu coal gasification, in which about 99.7 percent of the sulfur in the coal could be removed before combustion.

For this comparative study, the low-Btu coal gasification, combined cycle plant has been selected as the representative advanced fossil system that will be available in the 1990 to 2000 time frame. Using the same economic approach as used for the solar power plants, the total construction costs in 1975 dollars for a hypothetical 1975 plant startup were found to be $470 per kilowatt (electric) for the light water reactor and $454 per kilowatt (electric) for the coal gasification plant using wet cooling towers.[14] Dry towers would increase the capital costs to $589 and $540 per kilowatt (electric) respectively, and reduce plant efficiency to about 90 percent of wet. The energy cost with wet towers would be 22 mills per kilowatt hour (electric) for the light water reactor, and 32 mills per kilowatt hour (electric) for the coal plant, assuming coal costs $0.89 per million Btu ($23 per ton) and that the plants produce electricity, on the average, at 70 percent of their rated capacity. These energy costs are levelized over a 30-year plant life and include differential inflation for labor, material and fuel. Thus, the central receiver solar plant, if it were starting up today, would be 2 to 4 times the capital and energy cost of these nuclear and coal plants.

Although solar plants will be available starting 1985 to 1990, they will not be producing significant power before the year 2000. Moreover, the trends in nuclear and coal plant costs in the last decade have been consistently above general price inflation. Thus, to compare solar with coal and nuclear today is not particularly meaningful. A major question becomes, how long will these accelerated costs for coal and nuclear plants continue—

1 year, 1 decade, or what? In a sense, this extraordinary growth in conventional power plant costs is society's internalizing what heretofore have been external costs.

## Nuclear Costs

A light water reactor that was estimated to cost $140 per kilowatt (electric) in 1967 for a 1972 startup was estimated to cost $720 per kilowatt (electric) in 1973 for a 1983 startup. And today, this estimate is closer to $1,300 per kilowatt (electric). The rate of growth of nuclear power plant capital cost has been 16 percent to 20 percent per year over the last decade. The average inflation of installed capital goods was closer to 6 to 8 percent during this same time period. This amounts to about 10 percent differential inflation.

Coal plants were not far behind at a 16 percent per year capital cost increase[15] and a doubling in the average cost of coal to the utility industry in just one year (1973 to 1974).[16] If one plant is inflating at the general price inflation rate and another at 10 percent more, their relative cost difference will halve in about 7 years. Thus, a capital cost difference of 2 to 4 times today may not mean solar plants will not be competitive tomorrow.

To illustrate this point, a projection is made of plant construction costs for solar, coal, and nuclear versus time until the year 2000. The solar plant is the central receiver type with dry cooling with six hours of storage while both the light water reactor and the coal-fired plant are considered to use wet cooling. This advantage of wet cooling is given to the conventional plants since they may have somewhat more flexibility in siting choice. If dry cooling is needed in the year 2000, conventional capital costs should be increased by 25 percent for nuclear and 19 percent for coal. Also conventional fuel costs should increase by 11 percent for dry cooling.

If a solar plant is considered socially acceptable compared to other choices, its capital plant costs and plant operation and maintenance should inflate at the rates shown in table 5.2. Although solar plants are relatively clean with modest social impacts, it is unlikely that they will be embraced by all Americans as totally acceptable possibly because of the large amount of land used at the plant.

Solar plants may also not be immune from cost escalation due to social resistance. But this resistance, if any, should not develop until significant introduction of these plants has occurred. This probably will not happen until after the year 2000. For a variety of reasons, however, nuclear and

Richard S. Caputo

**Table 5.2**
Projected inflation rates for 1978–2000

| | Inflation rate in | |
| --- | --- | --- |
| | 1977–1985 | 1986–2000 |
| General price level | 5.0 | 4.2 |
| Manufactured goods | 4.3 | 3.8 |
| Construction labor | 7.0 | 6.2 |
| Plant operation and maintenance | 6.3 | 5.6 |
| Installed capital | 6.2 | 4.8 |

coal plants are suffering today from broad social resistance as a result of society's attempts to make them more acceptable.

An estimate of the future nature of this effect, with a range of future capital plant escalation, is shown in figure 5.6. The upper curves represent a gradual transition from today's high escalation rates to long-term rates by the year 2000. For the lower curves it is assumed that the long-term escalation rates are in effect today and continue for the rest of the century. The escalation rates used are shown in table 5.3.

With these escalation rates, solar and nuclear capital costs become more comparable by the end of the century, while the solar plant capital cost is still about one-half more than that for a likely coal plant. When fuel costs are taken into account as well, the cost of electricity from plants coming on-line in 2000 would be similar for all three system types. The levelized energy cost of solar, nuclear, and coal plants (over the 30-year plant life) is shown in table 5.4.

These projections, although not rigorous, appear to be a reasonable extension of present trends. This exercise shows that the significant cost differences that exist today between solar and conventional power plants may not necessarily hold up by the time large-scale commercial-size solar power plants are producing power in the year 2000.

## Social Costs

Besides capital and energy costs, there are a number of other factors that must be considered in comparing various power plant options. These factors are the research and development investments to create the first commercial version of these advanced plants, thermal pollution, land requirements, and health effects.

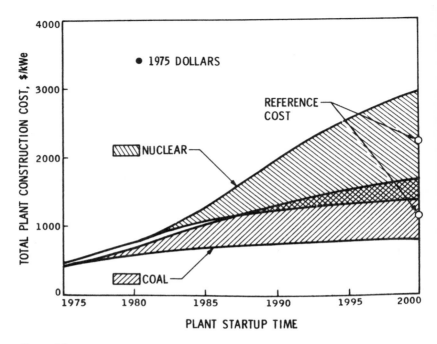

**Figure 5.6**
Projections of conventional plant capital cost. The Upper bound of the nuclear and coal curves represent a gradual transition from today's high escalation rates for these plants to long-term rates by year 2000; lower curves represent an immediate adoption of long-term escalation rates for conventional plants now. Given these projected escalation rates, solar and nuclear capital costs will become comparable by the end of the century.

**Table 5.3**
Plant capital cost differential escalation factors[a] (percent)

| Type | 1975–80 | 1980–85 | 1985–90 | 1990–95 | 1995–2000 |
|---|---|---|---|---|---|
| Nuclear | | | | | |
| Lower | 10 | 5.6 | 1.2 | 0.6 | 0.6 |
| Mid | 10 | 8 | 6 | 4 | 2 |
| Upper | 10 | 8.75 | 7.5 | 6.25 | 5.0 |
| Coal | | | | | |
| Lower | 4.25 | 2.4 | 0.6 | 0.6 | 0.6 |
| Mid | 4.25 | 3.3 | 2.4 | 1.5 | 0.6 |
| Upper | 8.5 | 6.8 | 6.5 | 3.4 | 1.7 |
| General price inflation | 5 | 5 | 4.2 | 4.2 | 4.2 |

Note: Total inflation rate equals general price inflation plus differential escalation.
[a] Fuel cost differential escalation from 1975 to 2000:
   Coal: Lower = 1%, Mid = 2%, Upper = 3%.
   Nuclear: See text.

**Table 5.4**
Levelized energy costs of solar, nuclear, and coal plants over the 30-year plant life
(1975 dollars and 8 percent interest on capital)

| | Plant start-up (mills/kWe) | |
|---|---|---|
| | 1975 | 2000 |
| Solar | 85 | 104 |
| Nuclear | 22 | 76 |
| Coal | 32 | 58 |

Solar Power Plants

The federal research and development cost for coal technology considered here (low-Btu gasification and combined cycle conversion) is about $1.5 billion. Future research and development for the light water reactor is estimated to be about $1.5 billion, while the breeder reactor program is estimated to cost more than $10 billion.[17] Federal research and development for the solar-thermal-electric plant (power tower) is estimated to cost $1.1 billion.[18] These costs represent the amount to be spent from the present to the date of commercialization.

Based on preliminary study, the solar plant is relatively neutral as far as excess heat rejection is concerned: that is, heat left at the site is similar to what would have been deposited if no plant were there. The central plant leaves behind an excess of about 0.25 megawatts (thermal) for each electric megawatt generated. This should be compared to 2.1 megawatts (thermal) per megawatt (electric) and 1.7 megawatts (thermal) per megawatt (electric) for a light water reactor and coal plant.[19] For the most part, the solar plant is using energy that would have been absorbed by the soil and local atmosphere anyway.

It is commonly stated that a major drawback of solar energy is the extensive land-use requirements. However, the amount of land required for a solar plant is actually comparable to what is required by the system which provides electricity from coal. To see this, note that the 3.2 square kilometers required for a 100 megawatt (electric) plant (see table 5.1) amounts to 2,000 square meters per megawatt (electric) per year, averaged over the 30-year output of the plant.

On the same basis a coal plant would require about 3,000 square meters per megawatt (electric) per year, if the land required for coal mining is taken into account. While the area requirements are comparable in these two cases, operation of the solar unit would actually involve a more benign use of the land.

Quantifiable health effects due to the systems considered here for providing electricity based on coal, nuclear, and solar energy are shown in table 5.5. The solar estimates are based on the pollutants generated from uncontrolled primary material processing plants used to manufacture all the materials needed for a solar plant, that is, steel, aluminum, concrete, glass, silver, etc., and plant materials acquisition and plant construction.

Based on this analysis, a solar plant has less health impact than a nuclear plant. The solar plant is about a factor of 40 better than the advanced coal plant with 99.7 percent control of the sulfur oxides. (This figure does not

Richard S. Caputo

**Table 5.5**
Comparison of coal, nuclear, and solar health impacts

| Impact area | Person days lost per megawatt (electric) year | | |
| --- | --- | --- | --- |
| | Coal[a] | Light water reactor | Solar |
| Occupational[b] | | | |
|   Accidents[c] | 19–58 | 1.8–3.2 | 5.8 |
|   Disease | 0.03–0.4 | 0.2–1.0 | 0.06 |
| Public, routine | | | |
|   Accidents[c] | 4.5 | 0.08 | 0–?[d] |
|   Disease[e,f] | 0.2–133 | 0.5–1.1 | 0–1.0 |
| Public, large accidents[c,h] | — | 0.003–10.7 | — |
| Total | 24–201 | 3.4–17 | — |
| Total deaths per 1,000 MWe plant over 30 yrs | 71–530 | 8.6–51 | 7.7 |

Source: K. R. Smith, J. Weyant, and J. Holdren, "Evaluation of Conventional Power Systems, Energy and Resources Program," ERG–75–5 (Berkeley, Ca.: University of California, July 1975).

[a] Low Btu gasification with combined cycle, assumes no pollutants at gasification plant.
[b] Does not include occupational hazards of plant material acquisition, fabrication, and plant construction.
[c] 6,000 persons days lost (PDL) per death, and about 50 PDL per injury.
[d] Small, but unknown at this time.
[e] The coal derived public diseases from sulphur oxides and particulates only; nuclear and long-term coal wastes ignored.
[f] Air pollution causes premature deaths, while radiation causes cancer deaths.
[g] No controls at primary materials plant.
[h] Nuclear deaths based on draft version of WASH–1400 (Rasmussen report); High end modified by a factor of 17 for latent cancers, based on report to American Physical Society by their Study Group on Light-Water Reactor Safety. Previous studies would extend range by about a factor of 6 in each direction.
[i] This does not include genetic effects, somatic illness, property damage, sabotage, and nuclear material diversion.

include the effects of pollutants besides the sulfur oxides and assumes no pollutants at the gasification plant.)

These comparisons probably underestimate the environmental benefits of the solar option. The coal calculations, for example, do not reflect the potential for climatic disruption from the continued burning of fossil fuels and the associated particulate emissions and the build up of atmospheric carbon dioxide. Moreover, the nuclear calculations do not take into account the risks of sabotage, diversion of nuclear materials to nuclear weapons use, and the problems of long-term waste disposal, all of which are difficult to quantify.

Moreover, the estimate of the impact of a large nuclear accident shown in table 5.5 should be treated cautiously. What is considered here is a very low probability event of very high potential consequence. Not only are the probabilities of such rare events very uncertain but also their full social impact is very difficult to estimate. One possible expectation is that a large nuclear accident would lead to a shutdown of the nuclear industry for months to years while investigations are conducted. In this sense nuclear power, like foreign oil, would be an insecure energy resource.

When compared to likely fossil and nuclear alternatives, the solar-thermal-electric plant can be reasonably competitive, assuming America's electric energy cost to be about 100 mills per kilowatt hour (electric) by the end of the century. This depends upon continued although decreasing social resistance for both fossil and nuclear plants, with general acceptance of solar. Predicting social attitudes a quarter century in advance is a risky business.

In addition to direct economic costs, there are a number of other characteristics of large power plants that should be considered when making a general social decision. A limited number were shown in table 5.6 and for the most part favor solar plants. And, as we pointed out above, these costs probably underestimate the advantage of solar energy because some major risks associated with coal and nuclear use are not readily quantified.[20]

Although a good deal of quantification is possible in a comparative study, the subjective judgments carry enormous weight in a socioeconomic decision.

• What value should be put on human life and health effects?
• What value should be put on low probability-high impact nuclear incidents compared to the relatively even and steady destruction of a coal plant?
• What value should we give to future generations if social instability exposes the environment to long-term nuclear wastes?

Richard S. Caputo

**Table 5.6**
Some central power plant impacts

| | Coal | Light water reactor | Solar |
|---|---|---|---|
| Health, safety | | | |
| ~People days lost per megawatt (electric) year | 23–109[a,b] | 2–14[b,c] | 0.5-1.6[b] |
| ~Max deaths per plant year[d] | 11.9 | 1.5 | ? |
| Federal research and development~ \$1 billion[e] | 1.45 | 1.40 | 1.13 |
| Commercialization date | 1984 | 1984 | 1990 |
| Excess waste heat~MWtyr/MWeyr | 1.7 | 2.1 | 0.25 |
| Land~m$^2$MWeyr | 3,000[e] | 100[f] | 2,000 |
| Water~1 million liters per MWeyr | 0.5[g]–9[h] | 1[g]–24[h] | 1[g] |
| Energy pay back~year[i] | 2.0 | 1.4 | 1.7 |

Note: MWtyr = megawatts (thermal) year; MWeyr = megawatts (electric) year.
[a] Considers only sulfur oxide-particulate pollution.
[b] Excludes power plant construction occupational impacts, and materials fabrication plant occupational impacts.
[c] Excludes genetic effects, illness, sabotage, weapon material diversion, long-term waste effects.
[d] Plant year = 700 megawatts (electric) year.
[e] Averages eastern deep-mined coal and western strip-mined coal.
[f] This will increase dramatically toward the end of the century as current high-grade uranium is depleted.
[g] Dry cooling towers.
[h] Wet cooling towers.
[i] Includes energy needed for construction materials as well as ancillary energy required in fuel cycle over 30-year plant life.

Solar Power Plants

- What value should be given to the potential diversion of weapon materials from the nuclear fuel cycle and their use for antisocial purposes?
- What value should be given to the use of southwestern lands for power production for the entire nation?

In the final analysis, choosing the mix of technologies for future power production is a social decision and needs broad input from throughout society so that we have some assurance that the systems coming on-line 15 to 30 years from now will be socially acceptable.

## Notes

1. Pumped hydro is where water is pumped from a lower to a higher elevation when excess energy is available, and allows it to fall and spin a water turbine when power is needed.

2. Aerospace Corporation, "Solar Thermal Conversion Mission Analysis, Midterm Report," Sept. 1975.

3. T. Fujita, "Preliminary Studies of Electric and Hydrogen Pipeline Distribution and Cost Optimization of Electric Transmission Lines," JPL 900-734 (Pasadena, Ca.: Jet Propulsion Laboratory, March, 1976). (The JPL 900 series of documents are internal documents.)

4. Aerospace Corporation, "Solar Thermal Conversion Mission Analysis, Semiannual Review," April 1975.

5. ERDA, "National Solar Energy Research, Development and Demonstration Program," ERDA-49, June 1975.

6. G. Kaplon, "Solar Thermal Energy Conversion Central Receiver and Test Facilities Projects Program Summary," paper presented at ERDA's semiannual review in Las Vegas, Nevada, Nov. 1975.

7. Large steam power plants are based on the Rankine thermodynamic cycle where heat is added to a pressurized liquid causing it to evaporate to the vapor state. This pressurized vapor is expanded through a turbine producing work (spinning shaft) and is then condensed at a low pressure by rejecting heat to the environment. Organic fluids such as freons or ammonia are used in a Rankine cycle at lower temperatures.

8. Energy transport via chemicals is achieved by using the heat at each parabolic dish to thermally dissociate a chemical such a methane into carbon monoxide and hydrogen. These gases are then transported at ambient temperature to a central plant where heat is released when the gases are catalytically recombined in a reactor. The heat is then used to generate steam to drive an energy conversion plant. The two sets of chemicals (recombined and dissociated) may be storable in underground storage volumes such as depleted oil and gas wells, aquifers, or mined caverns. The costs of these storage techniques is one order of magnitude less than other techniques. This would allow cost effective storage for days rather than hours. There are several development areas in this scheme.

9. M. K. Selçuk, "Survey of Several Central Receiver Solar Thermal Power Plant De-

Richard S. Caputo

sign Concepts," JPL 900-712 (Pasadena, Ca.: Jet Propulsion Laboratory, Aug. 1975). (The JPL 900 series of documents are internal documents.)

10. See n. 8.

11. In a Stirling cycle heat is added at constant temperature and the resulting expansion drives a work piston. Heat is removed from the gas at constant volume and regeneratively added after heat is rejected to the ambient. This heat rejection is at constant temperature while the gas is compressed.

12. A Brayton heat engine is the thermodynamic cycle used in jet engines and gas turbine engines. A gas is compressed (pressurized), heated at constant pressure and then expanded through a turbine producing work. If it is open cycle, the gas from the turbine is rejected to the environment. If it is closed cycle, the gas is cooled by rejecting heat and then starts the cycle again. Note there is only a gaseous phase for the fluid used in this cycle.

13. Because much higher temperatures would be involved with these heat engines than with the central receiver system, thermal energy would be difficult. Thus, mechanical, electrical, or chemical storage would be required (flywheels, pumped hydro, underground compressed air, batteries, or even hydrogen produced via electrolysis).

14. K. R. Smith, J. Weyant and J. Holdren, "Evaluation of Conventional Power Systems, Energy and Resources Program," ERG 75-5 (Berkeley, Ca.: University of California, July 1975).

15. Atomic Energy Commission, "Power Plant Costs Current Trends and Sensitivity to Economic Parameters," WASH-1345, Oct. 1974.

16. See n. 14.

17. See n. 14.

18. NSF/NASA Solar Energy Panel, "An Assessment of Solar Energy as a National Resource," Dec. 1972.

19. See n. 14.

20. The final report of this material has now been completed. See R. S. Caputo, "An Initial Comparative Assessment of Orbital and Terrestrial Central Power Systems," Final Report, JPL-900-780 (Pasadena, Ca.: Jet Propulsion Lab., March 1977.) (The JPL 900 series of documents are internal documents.)

Solar Power Plants

The tropical oceans collect and store tremendous quantities of solar energy. Recovery of this energy in a useful form could provide a major share of the energy for the United States and other industrial countries close to tropical seas. OTEC (Ocean Thermal Energy Conversion) is the term used by the U.S. Department of Energy (DOE) to describe its program for transforming into electric power the solar energy collected by the tropical seas.

An OTEC plant could operate much like a conventional power plant in the most widely discussed design concept, illustrated in figure 6.1. The heat input in a fossil fuel plant is just air heated by burning fuel; in an OTEC plant, the heat input is the surface water of the tropical oceans heated by the sun. Cooling in a fossil fuel plant is provided by rivers, lakes or coastal water; in an OTEC plant, the cold water comes from ocean depths of 2,000 to 3,000 feet. A cutaway view of a conceptualization of such a plant is shown in figure 6.2.

A plant of the size shown in figure 6.2 (300 feet in diameter) would have a capacity of about 100 megawatts, only one-tenth the capacity of a modern nuclear or fossil fuel power plant. More power would be obtained by having more plants rather than larger plants. Such plants would have to be spaced about 10 miles apart in a limited region so surface waters would not be cooled off too much. To give an idea of the resource base, such a network over the entire tropical ocean could supply the world's population in the year 2000 with the energy per capita now consumed in the United States. What fraction of this resource base can actually be harnessed, of course, depends on many factors—costs, potential alternative uses of the ocean surface, and environmental impacts.

While the operations of individual OTEC plants are likely to be relatively clean (unlike fossil fuel plants) no one knows what the climatic impacts would be of very large-scale OTEC development. Such problems should be thoroughly investigated as the first OTEC plants are being deployed.

In the introductory article to this collection, Frank von Hippel and Robert Williams point out that the capital costs of collectors and energy storage systems have been the major obstacles to the large-scale use of solar energy.[1] These economic obstacles are particularly serious when one attempts to convert solar radiation into electric power. A major attraction of the OTEC concept is that these obstacles disappear when the surface water of the tropical oceans is used for both solar collection and energy storage.

# 6 Solar Sea Power
## Clarence Zener

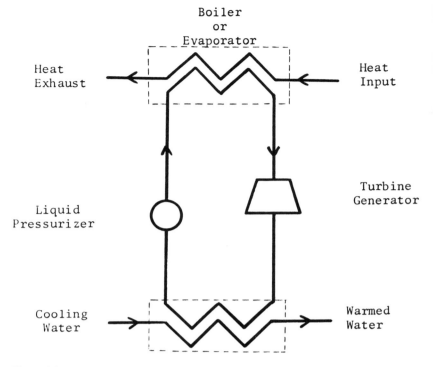

**Figure 6.1**
Basic essentials of OTEC and conventional power plants. Heat input is a hot gas in conventional plant, warm water in an OTEC plant.

But as von Hippel and Williams have pointed out, the OTEC system does introduce new costs, those associated with large boilers and condensers. As pointed out in this essay, these costs are no longer prohibitive.

The concept of using the ocean as a source of power has a long history. The French physicist Jacques d'Arsonval pointed out the possibility as early as 1881.[2] Georges Claude, a French chemist, actually built a demonstration plant off Cuba in 1930.[3] In the mid-1960s the Andersons pointed out the defects in the Claude design and made a fairly detailed cost estimate, which was based upon the concept of using very thin boiler and condenser walls allowed by precise balancing of internal pressures against external hydrostatic pressure.[4]

In the 1972–1973 period the National Science Foundation started

Clarence Zener

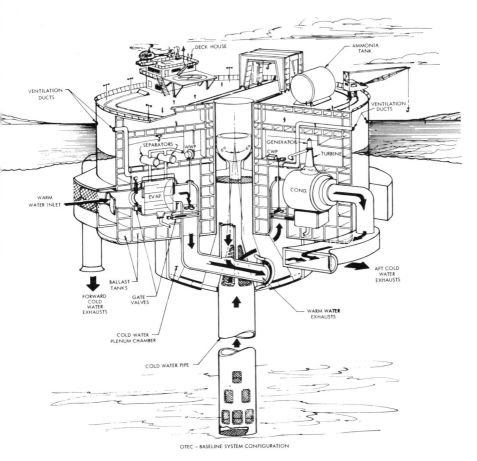

OTEC – BASELINE SYSTEM CONFIGURATION

**Figure 6.2**
Cutaway view of OTEC plant designed by TRW.[14]

Solar Sea Power

funding projects in this area. The first grant was to William Heronemus at the University of Massachusetts. The unique feature of his design was to use the strong current of the Gulf Stream to pump warm water through the boiler pipes. The second contract was to Carnegie-Mellon University. The particular design feature upon which the CMU group concentrated was an increased boiler and condenser effectiveness.[5-9]

Since this initial period the federal government has let many development contracts, culminating in contracts to Lockheed, TRW, and Westinghouse, leading toward the construction and operation of commercial size 100 MW OTEC plants by 1984.

The concept of an OTEC plant depicted in figure 6.1 employs a "working" fluid, such as ammonia, which circulates in a closed cycle—from boiler through turbine through condenser through pressurizer and back to the boiler. This concept is preferred by most OTEC workers because all steps have been thoroughly studied and have been found both technically feasible and economically attractive. Other OTEC workers find an alternative concept more attractive. The conventional concept depicted in figure 6.1 regards the upper layer of warm ocean water only as a source of heat. The alternative concept explicitly recognizes that an isentropic reduction in pressure plunges water into the two phase liquid-vapor region as depicted in figure 6.3. The area of the triangle ABC gives the ideal work which may be extracted in such a process. The embodiment of this concept into a practical OTEC plant requires maintaining a tight thermal and mechanical coupling between the liquid and vapor phases. Beck[10] is attempting to do this by an "air lift" device, Zener and Fetkovich[11,12] by foaming the warm sea water, Ridgeway[13] by misting the warm sea water. At the present early stage of OTEC development every concept which has a chance of becoming economically viable must be tested until we discover which is the most attractive.

### Typical Objections

During the past several years I have encountered a reluctance on the part of scientific friends to even consider the possibility of OTEC becoming economically viable. I believe that this reluctance stems in large part from a hope that the energy for the future will come from a system based upon highly sophisticated physics, certainly not from a system that at best could be called sophisticated plumbing. Because of this widespread feeling, I address below typical arguments against the concept of OTEC.

Clarence Zener

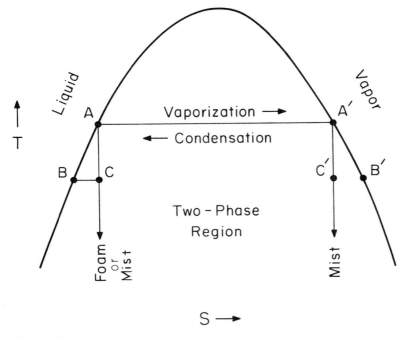

**Figure 6.3**
Typical temperature-entrophy (T-S) diagram.

*Because of the small temperature difference between the heat source and the heat sink, the efficiency will be very low. Very low efficiency implies a high cost per unit power output.*

The efficiency of OTEC plants will admittedly be very low, perhaps as low as 2 percent, in contrast to 38 percent for modern fossil fuel plants. However, efficiency per se is of importance only when we pay for the fuel. In our case the fuel is *free.* Our only concern is capital cost per unit of power output. The ideas highlighted in this essay show that this capital cost has a good chance of becoming comparable with that of competing conventional power plants.

*Because of the low efficiency of an OTEC plant, its boiler (that is, evaporator) must pass at least 15 times as much heat per unit net power generated as that of a conventional power plant. It therefore appears that the*

Solar Sea Power

*cost of the boiler of an OTEC plant must be at least 15 times that of a conventional power plant.*

The boiler tubes of a conventional power plant operate under the doubly difficult conditions of a high internal pressure, about 3,000 pounds per square inch, and a high temperature, about 1,000° F. In contrast, the boiler tubes of an OTEC plant operate under a nominal internal pressure of only 150 psi, and at ambient temperature. This difference in operating conditions means that the boiler of a conventional plant has no intrinsic cost advantage, per net power output, over that of an OTEC plant.

*No structure of the envisioned size of an OTEC plant could withstand the stresses induced by a stormy ocean.*

The ability of an OTEC plant to withstand storms has recently been investigated by Global Marine Development, Inc., of the TRW consortium.[41] They subjected a model of the TRW design to a worst-storm-in-100-years, and found it to withstand this test.

*The cooling water for conventional power plants is discharged at a temperature well above ambient. If OTEC plants were economically viable, they would be used as auxiliary power generators to conventional plants, being fueled by the warm discharge of the conventional plants. The observation that utilities do not use such auxiliary power generators is evidence of their uneconomic nature.*

Currently large heat exchangers are built on a job shop basis. OTEC plants would never be economically viable if their heat exchangers were to be built on the same basis. But the current rate of building new conventional stations does not encourage the development of automated manufacture. Once such production has been developed for OTEC plants, their use as auxiliary plants in central power stations may well become economically feasible.

*The requirement that OTEC plants are restricted to tropical waters implies that they can never become a major supplier of energy to continental United States.*

The United States is the only industrialized country in the northern hemisphere that is adjacent to areas of year-round warm surface water overlaying cold deep water. The warm water in the Caribbean flows northward into the Gulf of Mexico and then through the Florida Straits to become the Gulf Stream.

Clarence Zener

## Power Economics in 1980s

The future economic viability of OTEC will depend upon the cost of its power relative to the costs of power produced by conventional plants. To establish a benchmark for OTEC power costs, I shall therefore first review the costs of the more conventional alternatives. Because of the scarcity of natural gas, it is doubtful that any more gas-fired base-load plants will be built. Because of the nation's desire to become independent of foreign oil, it is also doubtful that any more oil plants will be built. The principal alternatives are therefore coal and nuclear plants, both of which are commonly regarded as our major energy supply options for the future. I shall use nuclear power costs as a basis for comparison because they are essentially independent of plant location.

The pertinent data for nuclear power plant costs are presented in figure 6.4.[15-17] The upper data refer to the average estimated costs of all plants contracted for during the year. The lower data refer to the average actual cost of all plants completed during the year. The costs refer to the total number of dollars paid out during the seven to eight years of construction, including the interest cost of the money already paid out. The straight lines representing the two sets of data have a slope corresponding to a 20 percent cost rise per year, that is, to a quadrupling in costs every seven years. Since at least seven years elapse from the time the decision to build a plant is announced to its completion date, we can expect a plant first planned for in 1977 will cost at least \$2,200/kw, namely, at least quadruple the average cost of plants first put into operation during 1977.

Our conclusion of a quadrupling of costs of recently ordered plants above the costs of plants recently put in operation is in marked variance with the relative position of the two straight lines of figure 6.4. The publicly announced predicted cost is less than twice the cost of plants just coming on line. This is because, as Olds has pointed out, the cost overruns on nuclear plants are typically over 100 percent of the original publicly announced costs. In 1975 Olds wrote, ". . . the industry is looking at 1985 costs of \$1,500–\$2,000/kw for plants it is ordering today at the same time new plants are coming on line for \$300–400/kw." I can only conclude that the nuclear industry has purposely withheld from the public, including their own stockholders, their true cost estimates of the nuclear power plants they plan to build.

Consideration of the reliability of large nuclear power plants further clouds any estimates of the future costs of nuclear power. Because capital

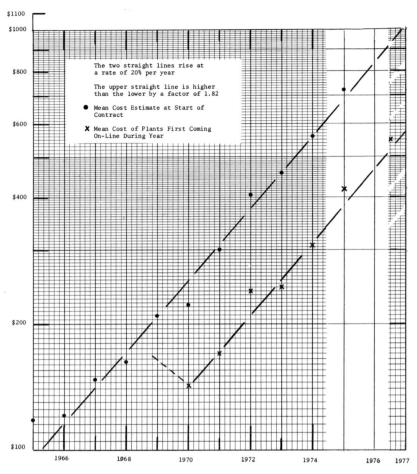

The two straight lines rise at
a rate of 20% per year

The upper straight line is higher
than the lower by a factor of 1.82

● Mean Cost Estimate at Start of
Contract

X Mean Cost of Plants First Coming
On-Line During Year

Figure 6.4
Nuclear power plant cost data.[16,17]

Clarence Zener

**Table 6.1**
TRW estimate of OTEC plant cost per net kilowatt output (1975 dollars)

|                                                                     | $/kilowatt |
| ------------------------------------------------------------------- | ---------- |
| Heat exchangers (evaporator and condenser)                          | 800        |
| Warm and cold water system (pumps, pipes, etc.)                     | 310        |
| Heat engine system minus heat exchangers (turbine generator, etc.)  | 210        |
| Structure and housing, engineering, etc.                            | 780        |
| Total                                                               | 2,100      |

Note: These figures are based upon current technology.

costs dominate the costs of nuclear power, utilities plan to use nuclear plants for base loading, that is, continuous full load operation apart from 20 percent downtime for maintenance. A measure of a base-load plant is its record on the ratio of number of kwh delivered over the year to the number of kwh that would have been delivered with continuous operation at full load. The utilities expect a capacity factor of 0.80. The average capacity factor[17] of nuclear plants greater than 750 MW capacity is 0.56. This unanticipated drop in capacity effectively raises the cost per kw capacity by 42 percent. Taking into account this low capacity factor, we must raise our estimate of the cost of nuclear power plants coming on line in the mid-1980s to an effective value of ~$3,000/kw. This capital cost represents a cost of ~60 mils/kwh, exclusive of fuel costs. We conclude that any renewable source of power that promises a capital cost of under $3,000/kw in the mid-1980s has a good chance of becoming economically viable.

Obtaining reliable cost data for untried processes is exceedingly difficult. Nonetheless the federal government has, through its agencies (first NSF, then ERDA, now DOE), made a strenuous effort to do just this for the OTEC power system. It asked Lockheed and TRW to develop conceptual designs independently using only proven technology and to estimate costs for a 100 MW plant. These costs vary essentially inversely as the square of the temperature difference between the surface and deep (~3,000 ft) water. This temperature difference is shown for typical locations in table 6.2. For a temperature difference of 39° F both estimates were about $1,800/kw in 1975 costs. Both companies expressed an opinion that these costs would be lowered to ~$1,600/kw by developments then under way on heat exchangers. The government then asked the National Academy of Engineering to review the Lockheed and TRW reports. As a result of this

Solar Sea Power

**Table 6.2**
Summary of average surface temperatures near the mainland United States

| | $T_s$ in °F | $\Delta T$, or $T_s - 40°F$ | $\Delta T$ penalty $(39°F/\Delta T)^2$ |
|---|---|---|---|
| On 3,000 feet contour off Gulf States | 76 | 36 | 1.17 |
| Southwest corner of Gulf of Mexico | 78 | 38 | 1.05 |
| Caribbean | 80 | 40 | 0.95 |
| Southern Pacific Coast of Mexico (Acapulco) | 85 | 45 | 0.75 |

Note: $T_s$ is the surface temperature; $\Delta T$ is the total temperature difference.

review, the government awarded contracts in the summer of 1977 to Lockheed, TRW, and Westinghouse aimed toward putting a 100 MW OTEC plant in the ocean by 1984.

Many people think it inappropriate to compare the economics of different power systems based merely upon today's price structure. They believe the increasing scarcity of energy will force the cost of energy intensive materials to rise faster than labor intensive materials. They therefore prefer to compare the total energy inputs rather than the estimated dollar costs. Toward this end the Institute for Energy Analysis[18] at Oak Ridge has made an exhaustive study of the energy input into an OTEC plant. They find that the electric power input would be repaid within six months of operation. The nonelectric energy input they regard as being a tax of 331 Btu for each kwh output. This tax is only one-thirtieth the typical 10,000 Btu heat input for each khw output of a fossil fuel plant. From the standpoint of energy input, OTEC is clearly far superior to either nuclear or fossil fuel plants.

### Future Role of OTEC in the U.S. Economy

The rapid pace of solar energy development in the United States insures that a major part of the future power we consume will be of solar origin. For our purposes it is immaterial whether the solar origin is direct, as for photovoltaics or heliostats, or indirect, as with wind or OTEC. Each source will find its particular niche. Wind power will dominate regions where the winds are strong and steady; direct solar power will dominate in the southwest where the sun is high and the air is clear; OTEC will dominate along those coasts bathed by warm ocean water. How far the solar generated power will extend will depend in part upon how low in cost the par-

Clarence Zener

ticular solar power system becomes, in part upon the overall development of low cost long distance power transmission.

Looking further into the future we can contemplate that OTEC plants based in the Gulf of Mexico will be feeding power into the Gulf states, that OTEC plants based in the Caribbean will stimulate energy intensive industry in the Caribbean region, and that OTEC plants based in the tropical oceans will furnish power to associated floating chemical plants for the production of basic chemicals such as ammonia, as discussed at length by the Applied Physics Laboratory of Johns Hopkins University.[19]

## Notes

1. Frank von Hippel and Robert H. Williams, "Solar Technologies," *Bulletin*, 31 (Nov. 1975), 25-31.

2. Jacques d'Arsonval, "Utilisation des forces naturelles: avenir de l' éctricité," *Revue Scientifique*, Sept. 17, 1881, pp. 370-372.

3. Georges Claude, "Power from the Tropical Seas," *Mechanical Engineer*, 52 (1930), 1039-1044.

4. J. Hilbert Anderson, James H. Anderson, "Thermal Power from Seawater," *Mechanical Engineer*, 88 (April 1966), 41-46; U.S. Patent No. 3-454-081.

5. C. Zener, "Solar Sea Power," *Physics Today*, 26:1 (1973), 48-53.

6. C. Zener, A. Lavi, "Drainage Systems for Condensation," Trans. ASME, 96(1) (July 1974), 209-215.

7. A. Lavi, C. Zener, "Plumbing the Ocean Depths," *Spectrum*, 10:10 (Oct. 1973), 22-27.

8. A. Lavi, C. Zener, "Solar Sea Power Plants—Electric Power from the Ocean Thermal Difference," *Naval Engineers' Journal*, 87:2 (April 1975), 33-46.

9. C. Zener, *Physics and the Energy Problem—1974* (New York: American Institute of Physics, 1974), pp. 412-419.

10. E. J. Beck, "Ocean Thermal Gradient Hydraulic Power Plant," *Science*, 189: 4199 (July 25, 1975), 293.

11. C. Zener, J. Fetkovich, "Foam Solar Sea Power Plant," *Science*, 189:4199 (July 25, 1975), 294-295.

12. C. Zener, "The OTEC Answer to OPEC: Solar Sea Power!" *Mechanical Engineering* 99:6 (June 1977), 26-29.

13. S. Ridgeway, "The Mist Flow OTEC Plant," Fourth Ocean Thermal Energy Conversion Conference, 1977, Paper No. 2-B-6.

14. TRW, "Final Ocean Thermal Energy Conversion Report to ERDA," June 1975.

15. F. C. Olds, "Power Plant Capital Costs Going Out of Sight," *Power Engineering*, 78 (Aug. 1974), 36-43.

16. F. C. Olds, "Environmental Cleanup 1975-1985: Huge New Costs, Little Benefit," *Power Engineering* (Sept. 1975), 38-45.

17. ERDA 77-30/3. "U.S. Central Station Nuclear Electric Generating Units: Significant Milestones." July, 1977, p. 12.

18. "Net Energy Analysis of Five Energy Systems," Institute for Energy Analysis, Oak Ridge Associate Universities, ORAU/IEA(R)-77-12, September, 1977.

19. G. L. Dugger et al., "Experimental Two-Phase Flow in Large Diameter Tubes of Heat Exchangers," Fourth Ocean Thermal Energy Conversion Conference, 1977, Paper No. 3-B-4.

Clarence Zener

The concept of producing electricity directly from sunlight is based on the photovoltaic effect, named after early observations that an electrical voltage is generated when light falls on suitable material structures. Although these early observations took place well over a hundred years ago, only in the last 20 years has major research and development work been done to develop photovoltaic solar energy conversion systems. Photovoltaic conversion devices, for example, proved themselves to be highly reliable power supply systems for satellites and space probes. This article is about photovoltaic conversion as a new terrestrial source of electric power generation.

The purpose of any solar energy utilization scheme is to provide an income energy for our activities as an alternative to relying on the capital energy resources (fossil fuel) now being consumed by man. Such a switch will be more readily accepted if it does not involve too great a deviation from our accustomed energy use patterns. In the last two centuries, with the development of an advanced technology based primarily on the heavy use of fossil fuels, man has been able to greatly reduce the dependence of his living habits on natural constraints, including the availability of sunlight. This sunlight availability is determined predictably by the earth's planetary motions and, less predictably, by meteorological effects within the earth's atmosphere. Efforts will thus be required to make at least a portion of the energy obtained from sunlight available at man's convenience.

Consequently, when any form of solar energy utilization is discussed, one can not talk simply about the use of an energy conversion device; a complete system includes—in addition to a multitude of solar cells assembled into a "solar array" to supply the desired voltage and current levels—a means by which energy can be stored and a means by which this energy can be released to the user in a regulated form. Such regulation will entail control of frequency and voltage to make its use compatible with our energy consuming devices.

Photovoltaic solar energy conversion systems transform light directly into electricity without the intermediate transformations to heat and mechanical energy characteristic of most other systems for electric power generation. This conversion process can take place in either a solid state device (solar cells) or in electrolytic devices, where it was discovered first. Three conditions have to be met to provide photovoltaic conversion:

● light must be absorbed by generation of mobile carriers of electrical charge, such as electrons;

---

# 7   Photovoltaic Solar Energy Conversion
# Martin Wolf

- a potential barrier must exist in adequate proximity to the generated mobile charge carriers to cause charge separation, the basic requirement for any electricity generator; and
- the charge carriers must be able to move easily through the system to its power output connections.

Although some versions of the electrolytic device could provide energy storage in addition to energy conversion, the solid state approach is currently favored, largely because higher efficiencies have been obtained. Additional features of the solid state system are: absence of moving parts or matter down to the atomic level; long life and little maintenance, if properly designed; operation essentially at environment temperature and, in most configurations, without special cooling provisions.

As there seem to be no significant enconomies of scale because of their modular nature, photovoltaic systems can easily be deployed at on-site locations, making use of the distributed nature of sunlight. They can be installed on rooftops to reduce interference with other land uses, and can also be designed to deliver a large part of the nonconverted energy as low temperature heat for space or process heating (sacrificing, however, some of the electrical conversion efficiency). Also, one of the most suitable materials for photovoltaic conversion is silicon, the most abundant element on the earth which is solid at normal use temperatures.

Given these attractive features, it has been envisioned by some that photovoltaic conversion will be the major method for the utilization of solar energy in the future. This should not be expected to occur in view of the large demand for low temperature heat (for space heating, hot water, process heating, etc.), which can be provided very efficiently by relatively simple methods of solar energy collection.[1] Nonetheless, it is quite likely that photovoltaic conversion will supply the major portion of the electric power to be obtained from solar energy. The basic technical feasibility has been well established through both the space program and numerous low power terrestrial applications, such as highway emergency call boxes, microwave repeater stations, warning lights on off-shore pumping platforms, etc. The major tasks yet to be accomplished involve system optimization (determining the life and performance of various systems), and the development of low cost mass production processes for both the converters (solar cells) and the energy storage means (batteries, fuel cells, flywheels, or similar devices).

The key element of the solid state solar electric system is the photovol-

Martin Wolf

**Figure 7.1**
Rooftop array of silicon photovoltaic cells atop the Mitre Corporation building in McLean, Virginia. The one-kilowatt solar system lights a bulletin board in the reception room.

taic converter, which is often pictured as a large assembly of rather small devices which are popularly called "solar cells." These converters can be fabricated from various semiconducting materials in several device configurations. The two major classes of semiconductors are those consisting of a single chemical element, such as silicon (Si), and those of two or more elements forming an inorganic compound, such as gallium arsenide (GaAs) and the copper sulfide $(Cu_2S)$/cadmium sulfide (CdS) combination. To date these semiconducting materials are the ones that have provided the most significant performance in photovoltaic energy conversion. But, then, why do development efforts continue to be spent on photovoltaic devices from numerous other materials as well as on these three? It is because each material does have different properties providing various advantages and disadvantages in system performance and ease of manufacture. But it is also a case where the old axiom still holds true: Things look the better, the less one knows about them!

Silicon, a nonmetalic element obtained from quartzite sand and the most abundant element next to oxygen in the earth's crust, is well explored as a semiconducting material and forms the basis of the current semiconductor device industry, including a modest production of solar cells. Silicon has also provided the highest solar cell efficiency achieved so far and an outstandingly stable device. Its disadvantage is a rather deep penetration of the longer wavelength photons, which requires the silicon layers to be of high crystal perfection and at least 50 microns thick to obtain good solar cell performance. Another disadvantage is silicon's relatively high chemical reactivity at high temperatures, which necessitates attention in the manufacturing process design.

Gallium arsenide had a long gestation period in the semiconductor industry, but is now extensively used for light emitting diodes. It has the advantages of providing a slightly higher theoretical maximum efficiency for solar energy conversion than silicon at 300° Kelvin. This performance margin increases with temperature, but seems to be significant only in operation above 100° Celsius, that is in a few specialty applications.[2] Further, gallium arsenide is theoretically useful in films approximately one micron thick. Its disadvantages are its high price and the limited availability of gallium.

Cadmium sulfide, a compound that occurs naturally as the mineral greenockite and that can be obtained as a precipitate by the action of hydrogen sulfide on solutions of cadmium salts, is currently being used in combination with a very thin layer of copper sulfide, which plays the major role in the conversion of photons to electricity.[3] The major advantage of this combination is the potential for extremely low-cost fabrication of this thin film device by evaporative and chemical processes. Its major disadvantages are its low efficiency, about half that of the silicon and gallium arsenide devices, and the necessity for design precautions to achieve device stability.

Gallium and cadmium are less readily available than silicon. Annual production in the United States is 2.3 tons of gallium, 4,000 tons of cadmium, and 120,000 tons of silicon. The world resource of gallium and cadmium is only about one-millionth that of silicon. Much of the gallium and cadmium is found in greatly diluted form; they are obtained primarily as byproducts in the extraction of aluminum from bauxite and zinc from zinc ores, respectively.[4] At the current state of development, silicon seems to

Martin Wolf

be ahead, with existing production facilities turning out 1,000 to 1,500 square meters of cell area annually. The majority of this production is now going toward terrestrial applications.

Both in silicon and gallium arsenide solar cells, efficiencies of 16 to 18 percent have been obtained in laboratory devices for terrestrial solar energy conversion,[5] while production line silicon cells exhibit 12 to 14 percent terrestrial solar energy conversion efficiencies at room temperature. The best efficiencies obtained from cadmium sulfide cells are 8 percent in laboratory cells, while efficiencies of cells from current pilot runs seem to be in the 5 to 6 percent range.[6] Although work is continuing on improving efficiency, it is not expected that it can be increased by more than one-fifth of its present best values in any of the three devices discussed. Such efficiency values may be perceived as low, but they are quite competitive with those expected from other methods for the conversion of sunlight to electricity, which involve the generation of heat and its conversion to electricity by more or less conventional routes. It may be noted that in order to obtain acceptable conversion efficiencies in photovoltaic cells, impurities in the materials used must be controlled during manufacture to a level of a few parts per million or better. Where chemical compounds are used as semiconductors, equally stringent standards on their composition must be maintained.

Solar cells have been produced routinely, although in small quantities, for nearly two decades. Devices for the storage of electrical energy, for the inversion from direct to alternating current and for voltage regulation have been produced in much greater quantities for a considerable variety of commercial applications. Thus, with the technology on hand, what is delaying a general switch to solar electric energy utilization?

## Cost of the System

The major problem is today's high price of solar cells. The solar cell prices are high because they are sold only in rather small quantities, and the large investment in process technology development and automation required to lower the price significantly does not appear warranted since the market has not been adequately developed. Thus, it will take government support to break this circle.

In a relatively free economic system, a new commodity source will not be able to replace an existing one unless it can compete with it in price. Consumers currently buy electricity, generated primarily from fossil fuels,

Photovoltaic Solar Energy Conversion

from the electric utilities at a price somewhere between 2 and 9 cents per kilowatt hour depending on location and type of use, but still mostly at a price in the lower half of this range.

Since the "fuel" for solar energy utilization is free, and since system maintenance costs are expected to be relatively low, the price determining component for solar electric energy is the capital cost of the system.[7] And at the current prices of $2,000 per square meter for silicon photovoltaic arrays, $30 per kilowatt hour of storage capacity for batteries, and $30 to $100 per kilowatt power rating for inverters and regulators, electrical energy by photovoltaic conversion would cost $.50 to $1.00 per kilowatt hour, depending on interest and other costs of capital. For general application solar cells are thus not competitive today.

There are exceptions, however. Numerous applications exist, involving mostly small power uses at remote locations, where solar photovoltaic energy conversion is competitive in today's market and where it is increasingly used. In these cases, the competing possibilities of installing a power line to supply the energy would involve greater capital costs, or of providing primary battery packs of a motor generator would necessitate frequent, costly trips for energy replacement and/or maintenance. While at present the cost of the solar cells dominates the photovoltaic system cost by a margin of close to 100 to 1, the costs of the other system components will be equally important once the prices of the solar cells have been brought down to acceptable levels to permit large-scale application.

## Optical Concentrators

While efforts are in progress to greatly reduce the solar array manufacturing costs, a parallel approach to cost reduction is being pursued: systems that use fewer solar cells.

In view of the fundamentally limited energy flux of the sunlight (a peak of about one kilowatt per square meter at the earth's surface), designing for a desired power output at a given converter efficiency requires the use of a corresponding receiver area. This receiver area does not have to be the converter; it may be an optical concentrating device. A considerable variety of such concentrating devices exists, providing either one-dimensional or two-dimensional concentration, with concentration ratios from close to one up to several thousand. The use of solar cells is then reduced by about the same ratio.[8] According to some estimates certain types of concentrators may be manufactured for under $10 per square meter. However, since

Martin Wolf

large-scale production experience does not yet exist, meaningful price/quantity or price/concentration ratio data are not available for these concentrators.

While nonconcentrating solar collectors can readily be used in a fixed orientation, systems with optical concentration require the capability of tracking the sun. Also, the optical perfection of the concentrator and the accuracy of the tracking system have to be higher the greater the concentration ratio. Thus, with rising concentration ratio, an increasing portion of the system cost shifts from the solar cells to the optical concentrating device. This leads to the question: Will we ultimately have a large solar cell industry, or a large optical concentrator industry? In either case, be it a flat solar array, that is, an assembly of numerous cells on a supporting substrate with protection from environmental influences, or be it a concentrator system with fewer solar cells and a tracking mount, the finished collector will have to cost less than $50 per square meter of collecting area.

Some additional disadvantages of concentrators are worth mentioning. As no system component performs in an ideal manner, concentrators also contribute losses that normally increase with increasing concentration ratio. In addition, the optical opening angle of the receiving system decreases with increasing concentration ratio. A small opening angle means that indirect sky radiation is received only from a narrow cone around the sun, while a flat collector accepts radiation from the entire hemisphere of the sky.

In many climates, the diffused sunlight radiated from the sky provides a significant portion of the total available solar radiation. Consequently, to satisfy a given electricity demand, the reduced solar energy utilization of a concentrating system has to be made up through a larger total collector area. This increased collector area is also a factor to be considered in cost computations. Thus, while the use of high concentration ratios appears beneficial at today's high solar cell prices, the optimal concentration ratio will decrease as the cost of solar cells declines. With the high energy density on the photovoltaic converter resulting from optical concentration, forced cooling becomes a requirement at concentration ratios above 5 to 10. Although not a technical problem, this cooling adds a cost margin and slightly reduces the net energy output.

## The New Industry

When we talk about large-scale terrestrial photovoltaic solar energy utilization, we think of replacing a significant part of the total energy supply by the new source. Thus, the magnitude of utilization of the new energy source is dictated by total energy consumption. Currently, electric power is consumed in the United States at a rate of approximately 2 trillion kilowatt hours per year. This, however, constitutes less than 10 percent of the total energy consumed by end users. Most current predictions assume that both the total energy consumption and the fraction used in the form of electric power will be significantly greater in the year 2020. These electric power generation forecasts, ranging up to 10 times the current values, will have to be treated with great caution, however.

To make an impact on the total energy situation, a significant fraction of the consumption has to be supplied from the new source. As an example, delivery of one trillion kilowatt hours per year from solar photovoltaic conversion systems would be about one-half of the current electric power generation in the United States, but only 5 percent of its total energy consumption. In the year 2020, the same one trillion kilowatt hours would provide between 5 to 50 percent of the electric power generation, depending on the specific energy forecast one is willing to give credence to. In any case, one trillion kilowatt hours per year would be a significant contribution.

This generation rate would require coverage with solar collectors of approximately 7,000 square kilometers of ground surface (see table 7.1). This appears as a very large number, but it constitutes only approximately 0.1 percent of the total U.S. land area, and about as much as has already been covered by buildings (see table 7.2). To install this capacity in a time period of 20 years would require an annual production rate of 350 square kilometers, approximately a million times higher than the current solar cell production rate. At a price of $20 per square meter this would amount to

Table 7.1
Estimates of potential production rate of solar arrays in the year 2020

| Energy impact viewpoint | |
| --- | --- |
| Goal: Total photovoltaic energy production rate in 2020 | $1 \times 10^{12}$ kWh/yr |
| At 10% efficiency and average daily insolation of 4 kWh/m$^2$, production rate = | 146 kWh/yr-m$^2$ |
| Total array area required (0.1% of U.S. land area) | $0.7 \times 10^{10}$ m$^2$ |
| Average production rate over 20-year period | $3.5 \times 10^8$ m$^2$/yr |

Source: "Data for Use in the Assessment of Energy Technologies," AET–8, April 1972. A report submitted to the Office of Science and Technology, Executive Office of the President, under Contract OST–30, Associated Universities, Inc., Upton, New York.

Martin Wolf

**Table 7.2**
Current U.S. land area uses

|  | Area (10 billion square meters) |
|---|---|
| Single-family residences | 0.39 |
| Highways | 5 |
| Cropland harvested | 110 |
| Total continental land area | 780 |
| Projected area for solar cell arrays | 0.7–1.0 |

an annual volume of $7 billion for the arrays alone, not counting storage, inversion, and installation costs. It is clear that such a huge increase in production rates, coupled with a decrease in price by two orders of magnitude, can occur only by a switch to an automated, continuous flow operation of much greater cost effectiveness than the one currently used. The array costs will always consist of material costs, labor costs, and capital costs for the manufacturing facility.[9]

Where the basic material costs are high, less material will have to be used. This may mean thin film approaches which can be practical for the gallium arsenide and cadmium sulfide devices. Fortunately, the raw silicon price cost is low ($1.00 per kilogram, comparable to the price of aluminum), so that in this case all efforts can be concentrated on reducing the manufacturing cost. Labor costs will normally be reduced by process simplification and automation. Automation, however, involves high capital costs for manufacturing equipment. To meet the cost goals for the solar arrays, the investment for equipment and plant will have to be less than $400,000 for each square meter per hour of array production capacity. This limitation generally leads to the demand for simple, high speed processes.

For the copper sulfide/cadmium sulfide thin film solar cell, a sequence of vacuum evaporation and chemical processes has been under development for the past 15 years. While some researchers felt that this process could lead to extremely low prices, other researchers convinced themselves that the unavoidable overspray in the vacuum evaporation process would prevent the attainment of adequately low costs. To alleviate this problem, development work is also in progress on a chemical spray method. This approach has been previously investigated with limited success, although the chemical spray approach has been routinely used in the cadmium sulfide photoconductor industry.[10]

Photovoltaic Solar Energy Conversion

The gallium arsenide device has the highest material costs per unit volume. To reduce costs, two approaches are being pursued: the use of single crystal wafers with optical concentrators of high concentration ratio; and a thin film design that would utilize either a polycrystalline gallium arsenide film on a low cost foreign substrate, or a thin single crystal gallium arsenide film grown on a substrate of matching crystal structure, but of a lower cost material.[11] While the concentrator approach is easier to accomplish with today's technology, relatively little has been done to develop low cost systems of either approach, beyond individual cell development.

In past efforts toward the understanding of the material properties and the device operation of semiconductors and toward the control of these parameters in production processes, most of the work performed and most of the progress made has been with silicon. The costs of the production processes for all types of silicon semiconductor devices, including solar cells, have been thoroughly analyzed, and detailed projections for future improvements have been made.[12] Much of the knowledge acquired with silicon can ultimately be extended to other semiconductor materials.

A study of the cost accumulation for solar array manufacture, starting from the raw material, quartzite sand, shows that costs are rather uniformly distributed throughout the process.[13] There is no single process step which makes such an outstandingly large cost contribution that its replacement could solve the cost reduction problem. Another remarkable factor is that the industry involved in the manufacture of semiconductor devices, particularly solar cells, is not vertically integrated.

The major process steps of quartzite reduction to metallurgical silicon, generation of trichlorosilane, its purification, subsequent reduction to high purity polycrystalline silicon, transformation into single crystal ingots, cutting wafers, and device manufacture are carried out by different firms. Various firms specialize in one or several but not all of the process steps. There are also device manufacturers who fabricate solar cells and sell them to other companies for assembly into arrays.

The reason for this structuring lies in the fact that the current semiconductor industry consumes only about 1 percent of the raw silicon production, while most of the trichlorosilane is used in the plastics industry. For the ultimate low cost fabrication of solar arrays in large scale, these conditions will have to change. And it may become essential to combine all the process steps into a unified sequence, hopefully in a continuous flow operation without much material transport between the process steps.

Until now, solar array production has been based on the manufacture

Martin Wolf

of individual cells with subsequent assembly on substrates. In this process, the individual cells are interconnected with special conductors in series to obtain an acceptably high voltage (as is done with electrolytic cells in batteries) or in parallel to increase the current output. In the effort toward cost reduction, the area of the individual cells has been gradually increased. Thus, for cadmium sulfide, 58 square centimeter cells (3 inches X 3 inches) are common; while for silicon, 16 square centimeter rectangular cells (2 X 8 centimeters) and 45 square centimeter circular cells (3 inch diameter) are now common.

The increase of the area of individual cells has the disadvantage of increasing the current output at the constant—low voltage of near 0.5 volt—thus further decreasing the already low internal resistance. This will gradually lead to a limitation in further increases of individual cell area. A more desirable development would be a shift toward "integrated arrays," which are similar in structure to the integrated circuits of modern electronics in so far as a multitude of components is contained and electrically interconnected within the same piece of silicon. This is, however, where the similarity ends.

In the integrated circuit industry, the cost reductions have been accomplished primarily through miniaturization. In photovoltaic array development, this approach is not feasible, since the limited energy flux from the sun makes large collector areas a fundamental requirement. Thus, it is necessary to learn to handle large quantities of material and to manufacture huge areas of arrays. This suggests the production of large sheets of semiconducting material, such as silicon, which incorporate a multitude of series connected cells to provide an output voltage compatible with the normal use voltages such as 117 volts.

In silicon solar cells, the optical properties of the material require, for good collection efficiency, the movement of charge carriers over distances of at least 50 microns.[14] This, in turn, requires a relatively high crystalline perfection. The exact degree of this perfection has not yet been determined, however.

Past experience has shown that single crystal material is not absolutely necessary for obtaining high performance in silicon solar cells. But it is not yet known what density of crystal boundaries can be tolerated. Frequently, the feeling is expressed that a diamater of 3 millimeters may be the lower limit for the size of the individual crystals in the polycrystal sheet. This limit will depend on the electrical properties of the crystal boundaries.

Photovoltaic Solar Energy Conversion

Figure 7.2
A continuous silicon ribbon for solar cells being pulled from molten silicon by the "edge-defined film growth" (EFG) process.

Martin Wolf

These properties will be influenced by the amount and type of impurity embedded in these boundaries.

A number of processes are under development which will avoid the current approach of pulling single crystal ingots of cylindrical shape from the melt (the Czochralski method), and the subsequent sawing of these ingots into wafers. This sawing process currently causes the loss of 50 to 70 percent of the valuable high purity single crystal material (as sawdust). Developments in progress hold forth the promise of considerable improvements in both ingot size and kerf losses in sawing, which may keep this process competitive with newer approaches.[15]

Processes under development to avoid the relatively costly steps of crystal growing and sawing are based on a continuous fabrication of ribbon or sheet of acceptable crystal quality. Such processes include chemical vapor deposition (CVD) that leads to polycrystalline sheets. These sheets can be recrystallized to single crystal or large crystalline material, for instance, by passing a narrow molten zone through the sheet.

Other approaches to silicon cell production utilize sheet crystal growth from the melt. One of these, the dendritic ribbon process, follows the previously explored route of pulling up a web of molten silicon between two needle crystals (dendrites) that grow spontaneously into a supercooled melt.[16] Thus, a ribbon of 1 to 4 centimeter width is obtained which contains generally three crystal boundaries parallel to the large ribbon surfaces. Silicon solar cells have, in the past, been fabricated from such dendritic ribbons with efficiencies as good as those obtained from Czochralski pulled wafers.

Another approach utilizes a die to shape a molten meniscus from which a ribbon is pulled.[17] This process is generally known as the "edge-defined film growth" (EFG) process that has found extensive use in sapphire production. The high reactivity of the silicon (molten silicon in contact with a wettable die material) has, so far, led to decreased electrical properties of the silicon ribbon grown by this method, resulting in solar cells with efficiencies near 8 percent.

Any one of the processes involving crystallization from molten silicon is limited in its speed by heat transfer considerations. Silicon has a high value of heat of fusion, which has to be removed from the solidification interface. Two interesting limitations result from these phenomena. First, the linear crystal growth velocity increases inversely with the thickness of the growing crystal, reaching 10 meters per hour in ribbons or sheets of 0.1 millimeter thickness. However, the volume rate of crystal growth decreases with decreasing cross-section of the growing ingot. To obtain the ultimately needed large volume of material at acceptably high speeds will require either growing large diameter ingots singly at low linear growth velocity, or pulling a multitude of ribbons or sheets simultaneously at high linear velocities. The latter approach presents problems in process control and equipment sophistication, which so far have not been solved. Numerous other processes are currently under investigation, including chemical vapor transport and high temperature plastic deformation (rolling) methods which may yield great process simplification and high speeds.

Once ribbons or sheets are obtained, the subsequent processes are more closely related to those currently used in the semiconductor industry. The only major difference is that they will have to be carried out on a large volume of material rather than on numerous small items of relatively high value. A large variety of processes are available, some of them tied to different device designs. How much costs can be reduced or what performance penalties may be incurred with the different approaches is still uncertain.

Finally, there has to be consideration of the life of a solar array. In order for a solar energy utilization system to be economically feasible, its operating lifetime will have to be adequately long. Twenty years without significant decrease in power output is usually considered a minimum lifetime. To achieve such a long lifetime, appropriate protective measures have to be taken. The copper sulfide/cadmium sulfide thin film solar cell, for instance, appears to require protection from both oxygen and water vapor.[18] Thus, a hermetically sealed encapsulation is required. A silicon device, on the other hand, appears to suffer only from electrolytic corrosion, as do all other conducting material combinations in the earth's environment. Thus, it needs an encapsulation that provides protection against excessive moisture, but hermetic sealing does not seem to be necessary. Good engineering design of the array structure will always be required, since wind forces and the considerable temperature variations, which arrays experience in the daily and seasonal cycles, can easily lead to damage to the mechanical structure and the encapsulation.

### Commercial Feasibility

It has become clear that none of the solar energy utilization approaches will become commercial without government support. This is especially true for the photovoltaic approach that will require a considerable investment in technology development and in sophisticated production equip-

Martin Wolf

ment to reduce the solar cell costs to the point where photovoltaic systems can produce energy competitively for the general market. But how should this support be provided?

Should the government support only research and development up to the point of proof of technical and economic feasibility and expect industry to then develop a commercial product on the basis of the know-how that thus has become publicly available? Or should the government support the industry with steadily increasing purchases of product, so that the industry can carry out its own proprietary process development and simultaneously acquire the production know-how and the manufacturing capacity to phase into the commercial market?

The second of these support schemes has been connected with the concept of the "learning curve," on which unit costs decline with increasing cumulative production of a specific device. The learning curve is a frequently used statistical tool in the electronic components industry, which usually has experienced a unit price reduction between 10 and 30 percent for each doubling of the cumulative output.

Extrapolation of the historical learning curve for silicon solar cells leads to adequately low prices for large scale terrestrial applications, using slopes corresponding to 20 to 30 percent price reduction for each doubling of production. The total price that would be paid to buy the entire product output would depend sensitively on the slope of the learning curve; it ranges from about $1 billion to well over $10 billion in order to reach price levels below $50 per square meter for slopes decreasing from 30 to 20 percent price reduction per doubling.[19]

The federal government has set 1985 as the target date for when solar arrays can be purchased at prices approaching those required to make large scale terrestrial applications feasible. Given today's knowledge and assuming both adequate financial support and superior program management, the goal appears reasonable. By that time, a few billion dollars will have been invested in the photovoltaic solar energy conversion approach, much less than expected to be required for the breeder reactor development. It also seems that the Energy Research and Development Administration has chosen an approach intermediate between the two extremes outlined above—supporting research and development and buying solar cells.

Assuming the development goals are met by 1985 and are followed by production at the rates discussed previously, it will be at least the year 2000 before enough power generating capacity can be built up to provide a significant contribution to the national energy supply from this new

Photovoltaic Solar Energy Conversion

source. The large expenditures, at the rate of $10 to $20 billion per year, will really occur during this period of capacity build-up. But these capital requirements are not significantly different from those needed for the build-up of a similar capacity for any other form of electric power generation. Such a statement can be made on the assumption that this capacity build-up will not be started until "competitiveness-condition" has been fulfilled. Any new energy technology that is not now fully production ready will require a comparable investment and a similar time before a corresponding impact on the energy supply can be experienced.

Thus, given that the basic technical feasibility for terrestrial photovoltaic solar energy conversion is well established, that the required materials are abundantly available, and that it is primarily a matter of substantial development to reduce sharply the costs of material and device processing, there are no basic obstacles to the achievement of large scale terrestrial photovoltaic solar energy utilization. From the year 2000 on, solar photovoltaic generated power could be of increasing importance to our national energy economy.

## Notes

1. W. A. Shurcliff, "Active-Type Solar Heating Systems for Houses: A Technology in Ferment," *Bulletin*, Feb. 1976; and R. W. Bliss, "Why Not Just Build the House Right in the First Place? [passive-type solar heating]," *Bulletin*, March 1976.

2. J. J. Loferski, *Acta Electronica*, 5 (1961), 350.

3. D. M. Perkins, *Advanced Energy Conversion*, 79 (1968), 265.

4. J. M. Woodall and H. J. Hovel, *Journal of Vacuum Science and Technology*, 12 (1975), 1000–1009.

5. R. J. Stirn and Y. M. Yeh, *Proceedings of the 11th Institute of Electrical and Electronics Engineers Photovoltaics Specialists Conference* (1975), pp. 437–438; M. Peckerar et al., paper presented at 1975 International Electronic Devices Meeting, Washington, D.C.; A. Muehlenberg et al., *COMSAT Technology Review* (1975); and L. W. James and R. L. Moon, *Proceedings of the 11th IEEE Photovoltaics Specialists Conference* (1975), pp. 402–408.

6. W. Palz, *Proceedings of the 11th IEEE Photovoltaics Specialists Conference* (1975), pp. 69–76.

7. M. Wolf, *Energy Conversion*, 14 (1975), 69.

8. C. E. Backus, *Journal of Vacuum Science and Technology*, 12 (1975), 1032–1041.

9. Wolf, *Journal of Vacuum Science and Technology*, 12 (1975), 984–999.

10. I. F. Jordan, *Proceedings of the 11th IEEE Photovoltaics Specialists Conference* (1975), pp. 508–513.

11. Woodall and Hovel, *Journal of Vacuum Science and Technology*.

Martin Wolf

12. M. Wolf, "Progress in New Low Cost Processing Methods," *Proceedings of the 11th IEEE Photovoltaics Specialists Conference* (1975).

13. C. G. Currin et al., *Proceedings of the 9th IEEE Photovoltaics Specialists Conference* (1972), pp. 363–369.

14. M. Wolf, *Energy Conversion*, 11 (1971), 63.

15. M. Wolf, *Proceedings of the 11th IEEE Photovoltaics Specialists Conference* (1975), pp. 306–314.

16. R. G. Seidenshicker et al., *Proceedings of the 11th IEEE Photovoltaics Specialists Conference* (1975), pp. 299–302.

17. K. V. Ravi, *Proceedings of the 11th IEEE Photovoltaics Specialists Conference* (1975), pp. 280–289.

18. J. Besson et al., *Proceedings of the 11th IEEE Photovoltaics Specialists Conference* (1975), pp. 468–475.

19. P. D. Maycock and G. F. Wakefield, *Proceedings of the 11th IEEE Photovoltaics Specialists Conference* (1975), pp. 252–255.

Most schemes proposed for converting solar energy into useful mechanical or electrical energy are based on relatively new technology. (See, for example, Clarence Zener, "Solar Sea Power," in this collection; and Martin Wolf, "Photovoltaic Solar Energy Conversion," in this collection.) In contrast, harnessing solar energy via the wind involves technology that is nearly as old as civilzation itself. Windmills were employed for grinding grains as far back as the seventh century in Iran and were important in bringing about the first mechanical revolution in Europe in the twelfth and thirteenth centuries.

Wind energy was harnessed early by man because in this form sunlight is already high quality mechanical energy. No heat engine is needed to make electricity from the wind; a simple windmill and generator can convert wind energy to electricity with high efficiency.

Wind energy is not used widely today largely because the wind blows only intermittently. Today our energy supply is dominated by sources for which output can be controlled easily to vary with demand.

Now that energy prices are rising and there is considerable interest in alternatives to fossil and nuclear fuels, wind power stands out as a solar technology requiring relatively little development before it can be producing economically considerable quantities of electricity. In this essay it is shown that wind power is economical at today's energy prices as a supplemental source of electricity, where wind power would be collected, converted to electricity, and transmitted directly to consumers via a utility grid system. Because no storage would be involved in such an arrangement, however, only 1 to 10 percent of the total electricity demand would be met this way.

To meet a much larger fraction of electricity demand, windmills must be used in conjunction with energy storage schemes that provide for consumer demand when the wind isn't blowing. In this case, wind power is not yet competitive with conventional power sources, and energy storage devices do require further development. However, even in this case wind power is close to being competitive. Moreover, the amount of research and development required on energy storage is far less than what is needed to make technologies like the liquid metal fast breeder reactor safe and competitive.

The atmospheric pressure systems that maintain the circulation in the entire atmosphere and oceans—driving the winds which, in turn, create the

# 8  Wind Energy
**Bent Sørensen**

waves—are generated by only about one percent of the radiant energy from the sun intercepted by the earth, amounting to 1,800 terawatts (or 1,800 trillion watts).[1] On average, this corresponds to 3.5 watts per square meter of the earth's surface (0.35 watts per square foot).

The flow of wind through a vertical square meter, say, at a height of 25 meters, however, may reach a yearly average of 500 watts per square meter at many windy locations, compared to the 3.5 watts mentioned above (see figure 8.2). A good windmill located at such a wind-favored location can convert to electricity about 175 watts per square meter of the area swept by its propeller. Given an appropriate configuration of an array of these windmills, it would be both technologically and economically feasible to feed electricity into an electric power transmission grid.

In determining an appropriate configuration for an array, one must investigate the relationship between the relative positions, heights, and the energy produced by a number of wind energy collectors, or windmills. Clearly, if the individual windmills within an array are placed very close to each other, there would be interference effects. Because of a reduction of wind speed through the individual windmill planes, the average energy content of the wind would diminish across the array—from the direction of the attack to the opposite end. Within the array, the wakes of individual wind mills can be regions of increased turbulence where there would be variations in wind intensity. This intensity could become substantially lower in some places and even higher in other places than the intensity of the original wind field.

From measurements behind fences, buildings, or woods, for example, one knows that the wind profile—the mean wind velocity as function of height—suffers a distortion which, under most atmospheric conditions, disappears after a distance of about 20 times the height of the obstacle.[2] Restoration of the wind flow behind a windmill requires a similar distance, though slightly less because the reduction in power will be smaller than for a solid object. The proper distance between windmills has been quoted by various authors as being in the range of 10 to 30 times the diameter of the rotor, depending on the height of the hub.[3] If a variation in wind intensity occurs, the mechanism for restoring the wind profile involves a mixing of the air which has been slowed down with the faster moving air masses above, so that the wind energy is fed back into the lower region.

A gross indication of which areas of the world are favorable for wind energy utilization may be gained from maps showing average wind speeds (see figure 8.3). Such maps are not really adequate for identifying favorable

Bent Sørensen

**Figure 8.1**
A 100-kilowatt experimental wind turbine atop a National Aeronautics and Space Administration tower at Sandusky, Ohio. The project is part of a U.S. effort to test the feasibility of wind energy systems for generation of commercial electric power. (Photo courtesy of NASA.)

Wind Energy

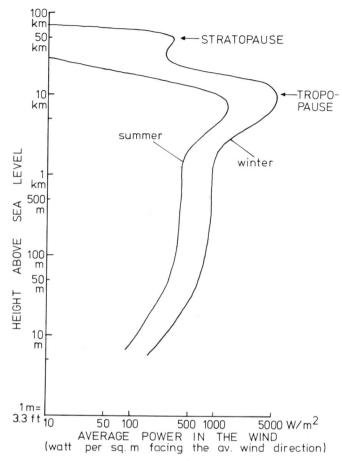

**Figure 8.2**
Variation of wind power with elevation (watts per square meter). The power in the wind increases with height, so that it may be feasible even where the ground wind speed is low to utilize wind power if the power is extracted at a sufficient height. The lower altitude data in the figure correspond to typical conditions in Western Europe and North Eastern America; the upper altitude data are common for areas of a northern latitude of about 55 degrees. *Source:* R. A. Craig, *The Upper Atmosphere: Meteorology and Physics,* International Geographical Series, Vol. 8 (New York: Academic Press, 1965). At higher altitudes, the fluctuations in the wind velocity are neglected in the power evaluation.

Bent Sørensen

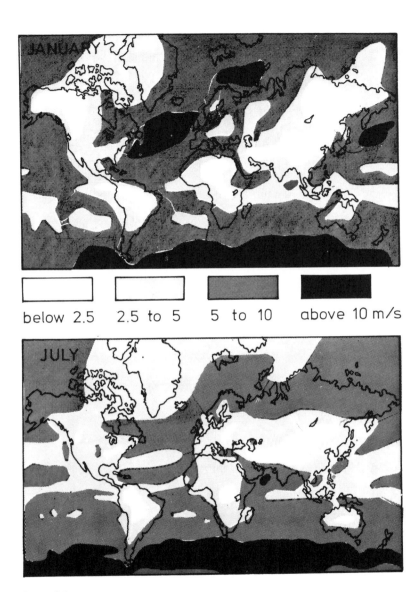

**Figure 8.3**
Global mean wind speeds (meters per second in January and July). This map, indicating which areas of the world are favorable for using wind power to generate electricity, is based on data taken at a height of 10 meters above ground. *Source:* World Meteorological Organization, *Proceedings of U.N. Conference on New Sources of Energy*, Rome, 1961, vol. 7.

Wind Energy

areas, however, because the average wind power can be considerably higher than the power developed by a wind of average speed and because within a given region wind conditions can vary considerably with local topography. The actual siting of windmills requires a detailed study of local topography. The classification made in figure 8.3 is based on fairly inhomogeneous data taken at a height of 10 meters. The general increase of power with height (curves of the type indicated in figure 8.2) may make it feasible to utilize wind power even in areas with low mean wind speeds if the energy is extracted at sufficient heights.

### Applications of Wind Energy

*Electricity Production* Wind energy is a high-quality form of mechanical energy that can be converted into electrical energy with minimal energy losses. This is significant because with most other systems for converting solar energy into electrical energy the greatest energy losses occur in the step when heat is converted into mechanical energy.

Since the rotor of a windmill moves periodically (typical frequencies are of the order of one revolution per second), the output may be obtained in the form of alternating current—either by using a gear-box and fixing the rotational speed or by allowing speed variations and transforming the generated electric power to the desired frequency, electronically.

Applications of wind energy range from small-scale use in isolated regions (including areas in developing regions without access to transmission lines) to large-scale generation of electricity which is fed into electric utility networks. In Europe both the wind power and the demand for electricity are greatest in winter. In the United States, while the wind power is greatest in winter, electricity consumption peaks during the summer in many areas. On a shorter time scale the amount of wind power generated does not generally correspond to demand, and a wind energy recovery system must be complemented by an energy storage facility or by auxiliary power from sources for which output can be varied to meet demand (fuels or hydropower, for example).

Some smoothing of the wind power can be expected if windmills at different locations are connected to the same transmission network. However, the mean fluctuation in the power output from a typical windmill is comparable to the mean power. The reduction in fluctuation of power output seems to be less than 10 percent for windmills located at a distance of up to 300 kilometers from each other since the wind regions change

Bent Sørensen

slowly, especially for the heights and wind speeds (typically above 5 meters per second) required for large windmills.[4] A somewhat larger reduction in the power fluctuation can be expected for windmills placed in entirely different wind climates, and attached to networks of corresponding sizes (extending, say, over thousands of kilometers).

The fluctuations above what would otherwise be the average power may be removed by using smaller generators, but clearly at the expense of a reduction in average power. The above estimates assume an electric generator that can handle about three times the average power output, as would be the case for a windmill with fixed rotational speed and blade profiles optimized with respect to yearly energy output.[5]

Another way to decrease power fluctuations is to add a short-time energy storage facility, with the capacity to deliver the average power for a period of time between some hours and a day or so. Such storage systems could be based on flywheels or compressed air located underground in excavated cavities such as salt domes or natural aquifers. Flywheels would seem preferable in connection with windmills because of the short time needed to shift from accumulation to generation and back.

Preliminary and rather uncertain estimates suggest that the cost of both flywheel and compressed air storage would be about $300 per kilowatt for each 10 hours of storage capacity, corresponding to about 20 percent of the windmill cost estimated below.[6] If a storage capacity corresponding to 24 hours of mean power generation is considered, the fluctuations in the resulting output could be reduced by about 25 percent, relative to a windmill installation without storage. In this case, the fraction of time the power output stays at the average, or above it, is increased from 40 to 70 percent, according to a calculation made for an inland site.[7]

If the storage cost estimates mentioned above hold up, it appears that short-time energy storage may provide advantages not easily obtained by the coupling of windmills at different sites. If wind energy should become a major energy source, long-term energy storage must be considered. The most promising method presently considered is the use of hydrogen, which will be discussed below.

*Heat Production* A heat-producing windmill can be built at lower cost than one producing electricity, since the electric generator as well as frequency and voltage control equipment can be replaced by an inexpensive heat generator, one based on generating turbulent motion of a fluid, for example. It is possible that such windmills could play a role in decentralized heating. On the other hand, such windmills would have to be sited where the heat is used, which would be in most cases at sites with less than optimal wind conditions for a given region.

However, conversion of a high-quality form of energy such as wind energy to a low-quality ("high-entropy") form of energy, as is needed for space heating, is inherently a wasteful process, and it should be avoided if reasonable alternatives exist. For heating purposes, rooftop solar collectors[8] or appropriately designed window systems[9] are preferable, and may be as economical as windpower, except at extreme latitudes.

An alternative is to combine direct solar and wind heating sources through the use of a wind driven heat pump which extracts low-grade solar heat from the environment (from a solar collector, outside air, a body of water, or the ground) and "pumps" this heat up to useful temperatures. When a heat pump is used with a solar collector as a heat source, a high solar collector efficiency is possible since the collector can be operated at a low temperature where heat losses are low. Also, typically less than half as much energy is required to produce a unit of heat, as with a direct heat generator, when the heat pump is used. The drawback of this approach is that it requires two capital intensive units—a windmill and a heat pump.

This combination seems to be uneconomical at present, except at a limited number of sites where wind conditions as well as the conditions for heat pump operation are especially favorable (for example, with lake or seawater as the heat source). Costs may also be reduced if the heat pump is run by electricity from a utility grid supplied with wind power instead of by on-site wind power. In this case capital savings may be realized by siting windmills at the most favorable locations.

*Hydrogen Production* An electricity-generating windmill may be used to produce hydrogen by electrolysis, when the electric generation exceeds the demand. The efficiency of the electrolysis process is today around 65 percent, but it is hoped that it can be increased substantially with efforts that will make this part of the conversion very cheap.

In one proposed scheme it is claimed that 93 percent efficiency might be obtained at a cost corresponding to about 5 percent of the price of the windmill. This would be for a system capable of converting half of the energy production to hydrogen by operating half of the time.[10] A further reduction in price, by a factor of 10 is hoped for by the year 2000.

The hydrogen which is produced may be stored or transported as a gas via pipelines, or as a liquid, or it may be trapped in lattice structures of

Bent Sørensen

metal compounds, with a decrease in volume but with an increase in cost. The hydrogen may be used for heating (buildings and industrial processes), propulsion and for regeneration of electricity (for example, by means of fuel cells—recombination of hydrogen and oxygen). At present, fuel cells have a short lifetime and an efficiency below 40 percent. For these reasons, the cost of regenerating electricity by way of hydrogen is still prohibitive.[11] But since this is an emerging technology, there is hope for economically acceptable solutions in the not-so-distant future.

*Other Uses* Largely because windmills are well suited for decentralized applications, windmills can be used for a number of special purposes at locations far from any energy distribution system. Such applications include pumping, irrigation, production of compressed air for later use, oxidation of lakes and so on.

By far the largest historic use of wind energy has been for propulsion, particularly of ships. A number of proposals have been made for reviving this use of the wind, with computerized control of the sails at a continuously optimized setting.[12] An auxiliary engine of modest size will ensure that such ships could be used as ocean freight carriers with travel times only slightly longer than totally fuel-driven bulk-carriers. Also, since the sailing vessel does not have to carry large engines and fuel, its freight tonnage may be larger than that of a comparable fuel-driven ship.

## Wind vs. Alternatives

In evaluating the feasibility of utilizing wind energy two distinct cases should be considered. First, the case where wind energy is envisaged as providing without storage only a small supplement to the conventional energy supply (say 1 to 10 percent). Second, the case where a massive introduction of wind energy is envisaged, requiring a substantial energy storage capacity. In the near future only the first case is of interest. But given the need for long-range energy planning, it is important to assess the second case, to provide a basis for comparison with other major energy alternatives.

If wind energy is considered only a supplement to a conventional energy supply, evaluation may be based on comparing the capital and operating costs of the wind energy system over its depreciation period, with that of the marginal production costs of current alternatives, that is, fuel and fuel-related costs. If wind energy is to be used on a much large scale, however, one must also ask how much other power generating capacity the wind energy system (including its energy storage facility) could replace. And in this case the comparison should be made between both capital and average operating costs of alternatives.

I shall present an example of such comparisons for windmills producing electricity which is fed into a large utility network. The cost of electricity (in constant 1974 dollars) is presented in table 8.1 for four alternative energy systems, along with the parameters entering into the calculations. The cost of electricity is taken as the busbar cost, averaged over a 25-year depreciation period.[13] Of course in practice the cost will vary during the 25-year period, according to the methods of financing and accounting.

Considerable uncertainty is attached to some of the parameter values—for example, those expressing expected future interest rates, wages and price of fuels and materials. In this sense the parameters of the table constitute reference values, and the effect of modifying some of the more important parameters will be looked at after the discussion of the table itself.

In columns two and three a comparison is made between the costs for a wind power system which is meant to save fuel only at conventional plants, and the costs of fuel for an oil-fired power plant. The windmills are supposed to be erected at windy locations and the capital cost estimate ($200 per square meter of swept area) is based on expectations for mass production of units of megawatt size, using the best of today's technology.[14] The updated capital cost of the 200 kilowatt Danish experimental mill at Gedser, which fed electricity into the public electric network during the period 1958–1967, is only 35 percent higher than this ($270 per square meter of swept area), suggesting that the cost we have used here is a realistic projection.

The resulting production cost of wind-based electricity is 1.3 cents per kilowatt-hour compared with 2.0 cents per kilowatt-hour for fossil fuels. In this case the difference is substantial. Thus, there is a little doubt that installation of supplementary wind power at wind-favored locations is economically feasible, and that it could help limit the demand for fossil-fuel resources.

In the last two columns of table 8.1, I compare a wind energy system, including an energy storage facility, with nuclear fission power. The capacity of the storage facility is such that the combined windmill-storage system generates the average power more than 80 percent of the time, if erected at favorable sites. In this way the power availability will be better than for

Bent Sørensen

**Table 8.1**
Comparison of electric power generation by different methods

| Parameters (reference values) | Windmill[a] | Oil[b] (fuel only) | Windmill and storage (crude estimate) | Fission[c] (light water reactor) |
|---|---|---|---|---|
| Capital cost | 200 $/m²[d] | 0.0 | 300 $/m² | 702 $/kW(e)[e] |
| Operation, maintenance | 0.3 ¢/kWh(e) | 0.05 ¢/kWh(e) | 0.3 ¢/kWh(e) | 0.14 ¢/KWh(e) |
| Average power production | 150 W/m² | — | 150 W/m² | 0.6 of installed kW(e)[f] |
| Fuel conversion efficiency[g] | — | 0.316[h] | — | 0.288 |
| Fuel cost | — | 0.60 ¢/kWh(t)[i] | — | 0.09 ¢/kWh(t) |
| Annual fuel cost increase[j] | — | 0.0% | — | 0.0% |
| Annual interest rate[j] | 3.0% | — | 3.0% | 3.0% |
| Construction period | 1 year | — | 1 year | 8 years |
| Depreciation period | 25 years | — | 25 years | 25 years |
| Annual increase in salaries and raw materials[j] | 2.0% | 2.0% | 2.0% | 2.0% |
| Calculated average cost of electricity[k] | 1.3 ¢/kWh(e) | 2.0 ¢/kWh(e) | 1.7 ¢/kWh(e) | 1.4 ¢/kWh(e) |

Note: $/m² = dollars per square meter; $/kW(e) = dollars per kilowatt of electric energy; ¢/kWh(e) = cents per kilowatt-hour of electric energy; ¢/kWh(t) = cents per kilowatt-hour of thermal (heat) energy; W/m² = watts per square meter. [a]Sørensen, "Vindkraft," Summary and Supplement from the Wind Energy Committee of the Danish Academy of Technical Sciences (1975) (summary report available in English), which also gives details of calculation; Rosen, Deabler and Hall, Intersociety Energy Conference paper. [b]Sørensen, "Vindkraft." [c]D. Rose, "Nuclear Eclectic Power," *Science*, 184:4134 (April 19, 1974), 351–359 (partially in 1981 prices). [d]Expressed as the cost per unit area swept by the windmill (based on data in Rosen, Deabler and Hall's paper, assuming the average power output to be a third of the rated power). [e]Probably a low estimate, compare text. [f]Based on data in David Comey's paper, "Will Idle Capacity Kill Nuclear Power," *Bulletin of Atomic Scientists*, Nov. 1974, pp. 23–28. [g]Includes a 5 percent in-plant transmission loss. [h]For marginal utilization, that is, the ratio of electric power output to thermal input for the last kilowatt-hour produced. [i]Corresponding to $9.59 per barrel. [j]Corrected for inflation. Thus an interest rate of 3 percent corresponds to a market interest rate of 13.3 percent, if the yearly inflation is 10 percent. [k]The calculated average cost of electricity over the entire depreciation period is expressed in fixed 1974 prices; the cost of distribution network is not included.

a single nuclear power plant, according to present experience,[15] and comparable to the combined performance of 2 to 5 nuclear plants. My guess that the cost of the energy storage capacity is 50 percent of the estimated cost of the windmill itself is admittedly crude, although in accordance with the estimated cost of flywheel energy accumulators mentioned earlier.[16]

According to this calculation the cost of wind power (with storage) is 1.7 cents per kilowatt-hour, and thus is higher than that of nuclear power, which is 1.4 cents per kilowatt-hour. Still, nuclear power comes out more expensive than wind power without storage. The capital costs of nuclear power plants, as well as of nuclear fuels, seem to be rapidly increasing,[17] so the values given in the table may be too optimistic. If the capital cost of nuclear power were $970 per kilowatt instead of $702, or if the nuclear fuel price were to increase by 3 to 4 percent per year faster than the present rate of inflation, then the cost of electricity produced by nuclear and wind power (including storage) would become equal.

In the comparison of wind power and nuclear power, the major additional uncertainties involve the operating costs. Is it true that windmills will demand more repairs and more manpower than nuclear reactors?

The maintenance cost of nuclear power given in the table may be too low in view of the increasingly strict demands to ensure safety. In addition, the costs of high and low level waste disposal, security measures to prevent sabotage and proliferation of fissile materials, as well as the costs of decommissioning reactors and fuel support facilities, may have been adequately incorporated in the cost estimates of the table.

In order to exhibit some of the sensitivity to various key parameters, figure 8.4 shows the dependence of break-even price on wind power density, conversion efficiency, future interest levels and fuel prices for the case in which wind power provides only a small supplement to the conventional energy supply. The break-even price is defined as the highest investment cost for which a new energy system will be competitive.

In this case the reference parameters lead to a break-even price of $360 (1974 level) per square meter swept by the rotor, that is considerably higher than the estimated cost of $200 per square meter shown in the table. This figure is also higher than the actual price of the 1958 experimental windmill erected at Gedser (expressed in 1974 dollars), even with the low efficiency of 23 percent then measured.

Since the break-even price increases the higher the efficiency (see figure

Bent Sørensen

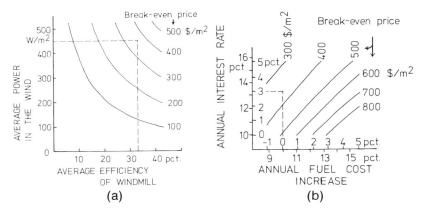

**Figure 8.4**
Break-even price of wind power. The break-even price is defined here as the highest investment price at which wind power will be competitive with the fuel costs of an oil-fired plant. Figure at left shows the dependence of the break-even price on location and overall efficiency of the windmill, taken as a fraction of the average power in the wind. The dashed lines correspond to the reference parameter values used in table 8.1, leading to a break-even price of $360 (1974 level) per square meter swept by the windmill. Figure at right shows the dependence of the break-even price on future interest rates and fuel prices, both corrected for inflation (numbers on the inside of the axes) or with 10 percent annual inflation (number on the outside of the axes). The dashed lines correspond to the reference parameters used in table 8.1.

8.4(a), the windmill constructor is motivated to seek the economically optimum efficiency: the extra cost of improving the efficiency is probably low when the efficiency is low, but steeply rises when it approaches the theoretical maximum efficiency of $^{16}/_{27} = 59$ percent. The dependence on the power of the wind shows that—at least with the present cost structure—windmills should be erected only at wind-favored locations.

Figure 8.4(b) shows that the economic advantage of windmills will increase rapidly if future fuel prices were to increase faster than inflation. This may indeed be the situation in the case of oil, due to the power of the oil companies to maintain profits during the period when easily extractable oil resources become depleted.

The oil price that I selected, and which is shown in table 8.1, is one which follows general inflation (in the western world), and it should be considered a conservative limit—one that is likely to be realized only if energy systems which do not use oil are rapidly developed. If fission power based on the nonbreeder reactors is to play this role, the assumption that

Wind Energy

nuclear fuel prices will not increase beyond the adjustment for inflation made in the table is unrealistic.

The investment necessary to ensure a yearly average production of one gigawatt (1,000 megawatts) of electric power is very high for wind power, about $1.3 billion for a system without storage. This figure is even higher than that estimated in the table for a nuclear plant with the same output, about $1.2 billion (including initial fuel). However, as the years of operation proceed, the accumulated fuel savings tend to compensate for the higher initial cost of the windmills.

On a life cycle cost basis the price of windmill energy is sufficiently close to that of energy from alternative sources that factors which are not easily quantifiable may be decisive in choosing between wind and alternative power sources. Governments, for example, could offset the high capital cost disadvantage of wind power by offering low interest loans for wind power investments. Such a policy would enable a society to make itself less vulnerable to the inflationary effect of fuel-based energy systems, and to the uncertainties of supply (particularly imports). It could also stimulate a transition to a less polluting source of energy.

### Environmental Impact

Environmental concern about windmill systems involves such factors as the large physical space required by wind-energy systems, compared to fuel-based conversion systems, the risk of accidents, noise, and the possibility of climatic alterations.

Since the ground area requirements are small and the land surrounding the windmills can be used with few restrictions, the space problem is mainly one of aesthetics. The accident risks mainly involve parts being expelled, for example, a blade breaking off during a storm. Such risks can be reduced by adequate design, strict inspections and restrictions on land use immediately surrounding the windmills. Other risks may be associated with the storage systems. One possible hazard is missile expulsion in flywheel breakdowns. Flywheels that decompose to dust in case of failure have been investigated.[18]

The noise problems are of two kinds. One is the noise created by the gear box and generator. This noise can be reduced far below the standards required, even in the most quiet conservation zones (35 decibels in Denmark).

The other type of noise is that created by the air flow off the rotor

Bent Sørensen

blades. It increases at least as fast as the power. It has been reported that with the tip velocities (the velocity of the fastest moving part of the blade) below 100 meters per second the audible noise of the wings is no problem.[19] This probably means that the noise is no greater than that of the wind itself, and in this case it would seem unreasonable to enforce, say, a 35 decibel noise limit in the immediate neighborhood of the windmill.

Since typical frequencies of windmill rotors are in the range of 1/2 to 3 revolutions per second, one may expect to find infrasound waves created in the wake of a mill. Intense infrasound waves may represent a biological hazard. However, the power carried by such waves, behind a windmill, will normally be very small and will be rapidly dispersed by turbulence. Thus, the possible concern about infrasound waves transmitted from a wind-power plant may be taken care of by placing restrictions on housing behind a windmill, at least until the problem has been analyzed in more depth.

Interference with TV and radio-frequency reception in proximity to windmills may be caused in part by the towers (like any other building structure) and in part by the rotors, depending on the materials used. Since the suitable sites for large, electricity-producing windmills are not expected to be those sites which are surrounded by housing, this is a matter which will probably be of concern only in connection with small windmills.

Another concern which has been discussed is the potential climatic impact of wind energy extraction on a large scale. While I find it difficult to envision significant effects on the general atmospheric circulation—on the basis of a preliminary estimation of the mechanism for restoring the wind profile described above and the data on energy generation and dissipation at different heights[20]—careful investigation of the possibility of climatic disturbances should be carried out before the wind conversion programs are implemented on a truly large scale.

## Summary

I have shown that among the possible uses of wind energy, electricity generated by windmills at favorable locations and fed into utility networks is technically feasible today. And within the limits of uncertainty, it is even economically feasible today, starting as a marginal supplement and eventually expanding to provide a sizable share of our electricity supply.

Economic feasibility requires that the evaluation be made over the entire lifetime of the energy systems. Uncertainties about lack of knowledge of future fuel prices, interest rates, etc., are relatively unimportant in this connection, because the estimated price of wind energy is, in any case, sufficiently close to that of energy from alternative sources to make the not directly economic factors decisive in the choice.

Such considerations include the uncertainty of fuel costs and supply, the adverse environmental impact of fuel-based energy production, and the growing understanding of the importance of restrained use of nonrenewable resources in the interest of future generations. Wind energy utilization is an attractive option for societies that are concerned about these constraints.

## Notes

1. E. N. Lorenz, *Nature and Theory of General Circulation of Atmosphere* (Geneva: World Meteorological Organization, 1967).

2. M. Jensen, "Shelter Effect, " unpublished dissertation, Danish Technical University, 1954.

3. M. J. Changery, "NSF/NOAA Initial Wind Energy Data Assessment Study," NSF-RA-N-75-020 (Washington, D.C.: National Science Foundation, 1975); R. Templin, paper presented at 1974 Workshop on Advanced Wind Energy Systems (Stockholm: Swedish Board for Technical Development, 1975); and O. Ljungström, speech presented at 1975 Energy Day at Danish Technical University.

4. C. G. Justus, appendix to "NSF/NOAA Wind Energy Study." I have adjusted the data in question so that the total wind power at the sites considered equals the sum of individual wind powers.

5. The use of one or possibly two fixed rotational velocities may ease the problem of resonance oscillations. This is a problem which could be expensive to solve if the rotational speed is allowed to vary freely.

6. R. Post and S. Post, "Flywheels," *Scientific American*, 229:6 (1973), 17; and G. Szego, "Energy Storage by Compressed Air," in *Wind Energy Conversion Systems*, edited by M. Savino, NSF-RA-W-73-006 (Washington, D.C.: National Science Foundation, 1973); Koordineret Kraftvaerksudbygning i 80' erne. Association of Danish Power Stations, Aug. 1975.

7. B. Sørensen, "The Case for Using Renewable Energy Sources," paper presented at 1976 Conference on Appropriate Technology for United Kingdom at Newcastle upon Tyne.

8. W. A. Shurcliff, "Active-Type Solar Heating Systems for Houses: A Technology in Ferment," *Bulletin*, Feb. 1976.

9. R. W. Bliss, "Why Not Just Build the House Right in the First Place?" *Bulletin*, March 1976.

10. W. Titterington, "Solid Polymer Electrolysis," in *Wind Energy Conversion Systems*, W. Hausz, "Use of Hydrogen and Hydrogen-Rich Components," in *Wind Energy Conversion Systems*.

Bent Sørensen

11. Titterington, in *Wind Energy Conversion Systems.*

12. Dynaship Corp. (Copenhagen), pric. comm.; J. King, discussion at 1976 Conference on Appropriate Technology.

13. Busbar cost is the cost of electricity at the generating plant. The price the consumer pays also includes the cost of transmission and distribution.

14. G. Rosen, H. Deabler and D. Hall, paper presented at 10th Intersociety Energy Conversion Engineering Conference in Newark, Del., Aug. 1975.

15. D. Comey, "Will Idle Capacity Kill Nuclear Power," *Bulletin,* Nov. 1974; H. Kohn, "Reactor Performance Evaluation," *Power Engineering* (Dec. 1975), p. 52.

16. Post and Post, *Scientific American.*

17. *Business Week,* Nov. 17, 1975, p. 98.

18. Post and Post, *Scientific American.*

19. M. Johansson, "Research Association of Danish Electricity Supply Undertakings," DEFU-TR-152 (1974); English translation available from NASA, report TT-F-16058.

20. E. Kung, *Monthly Weather Review,* 94 (1966), 627.

# IV  Fuels from Solar Energy

The major challenge in harnessing solar energy is to overcome two troublesome features of sunlight. Given that it is both intermittent and diffuse, sunlight is an elusive energy source. Because it is intermittent a solar energy storage system is needed; and because it is diffuse a large collector is required. Together these components account for a large fraction of the high capital costs of most solar conversion devices.

In nature the process by which green plants make use of sunlight (photosynthesis) is one in which these constraints are mitigated. With photosynthesis the collector cost is usually the cost of the needed land or water surface, and storage is conveniently provided in plant matter. This solar conversion process offers a renewable resource in especially convenient forms to replace diminishing supplies of oil and gas, since relatively simple technologies are required to convert organic matter to high quality fuels like methane, methanol, and hydrogen. Moreover, photosynthesis can provide for the long-term attractive, low polluting, renewable alternatives to coal as an industrial boiler fuel.

With a properly designed and managed photosynthetic conversion system, it is possible to keep adverse environmental impacts to relatively low levels. Raising plants as fuel provides a global pollution advantage over fossil fuels. Unlike the burning of fossil fuels, the burning of cultivated plants or fuels derived from them leads to no net build-up in atmospheric carbon dioxide,[1] since in a steady state carbon dioxide would be consumed in photosynthesis as fast as it would be released in combustion. Also the sulfur content of plant matter is generally below 0.1 percent (compared to an average of about 2.5 percent for coal), and the ash content of terrestrial plants is typically 3 to 5 percent (compared to an average of about 14 percent for coal).

A major limiting factor of photosynthesis, however, is the small efficiency involved. Nature converts only a fraction of a percent of incident sunlight to stored energy in plant matter. Even under especially favorable conditions photosynthesis converts no more than 1 to 4 percent of incident solar energy to chemical energy. Nevertheless, photosynthesis stores each year some 17 times as much energy in plant matter as that which is presently consumed worldwide.[2]

In the energy affluent United States, photosynthetic production on land is roughly comparable to total energy consumption. On the average, natural photosynthetic production on 7 acres yields the equivalent of per capita U.S. energy use (though in any attempt to harvest plant biomass not all the net primary production could be recovered). For comparison, the

# 9  Flower Power: Prospects for Photosynthetic Energy
## Alan D. Poole and Robert H. Williams

**Figures 9.1 and 9.2**
With nutrients from raw sewage, this water hyacinth plantation in Bay St. Louis, Miss., produces biomass, which can be converted, using a methane gas converter, into biogas for fuel.

amount of land required to provide a person in the United States with food is about one-third of this.

## Fundamentals

In photosynthesis green plants utilize sunlight to produce glucose ($C_6H_{12}O_6$), oxygen ($O_2$), and water ($H_2O$), out of carbon dioxide ($CO_2$), water, and an adequate supply of nutrients. In this process solar energy is stored as chemical energy in the glucose produced. Dry plant matter generally has an energy content of about 4 kilocalories per gram [3] (or about 15 million Btu per ton), which is about 60 percent of the energy content of bituminous coal.

Poole and Williams

The efficiency of converting solar energy to chemical energy in plant matter depends on various factors, some of which can be manipulated by man to optimize biomass yields. A small fraction of available sunlight is captured in photosynthesis. Only light with wavelengths between 4000 and 7000 angstroms is photosynthetically active. On a clear day this accounts for 40 to 45 percent of total sunlight. Furthermore, about 75 percent of the photosynthetically active radiation is lost through inefficient utilization of the energy delivered by light photons. Combining these two effects means that nearly 90 percent of natural sunlight cannot be converted to chemical energy in plants for fundamental reasons. Beyond this the photosynthetic conversion efficiency is reduced by losses that are less fundamental, such as reflection losses, physiological losses, inactive absorption, carbon dioxide limitation, and ground coverage losses. These losses are subject to limited manipulation by man. The practical maximum photosynthetic solar energy conversion efficiency that one might expect to achieve is on the order of 4 percent.

Flower Power

In typical situations the photosynthetic conversion efficiency is much smaller than the ideal (see table 9.1). The greatest photosynthetic production occurs along the equator. Though the equatorial belt is not the part of the earth where the greatest solar insolation occurs, the equator is where the best conditions for growth frequently occur—plenty of water and a year round growing season. (The rate at which solar energy strikes a unit of horizontal surface is called insolation. It includes both direct and diffuse solar radiation. The value of solar insolation at the earth's surface varies not only with the time of day and latitude but also with the time of year, cloudiness, and the dust and water vapor content of the atmosphere. Rates of precipitation tend to be higher in equatorial regions, as illustrated in figure 9.3. The most productive ecosystems within the area are the tropical and subtropical bogs.[4]

Table 9.1
Examples of photosynthetic production

|  | Annual average insolation watts per sq. ft. | Production Btu/acre/yr. (millions) | Efficiency (percent) |
| --- | --- | --- | --- |
| Theoretical maximum (global average) | 14 | 950 | 5.2 |
| Global average | 14 | 30 | 0.16 |
| Terrestrial systems |  |  |  |
| U.S. average (48 states) | 17 | 54 | 0.24 |
| U.S. agriculture (typical average values) | 17 | 55–160 | 0.25–0.75 |
| Eucalyptus, California | 19 | 230–410 | 0.94–1.7 |
| Exotic forage sorghum, Puerto Rico | 23 | 440 | 1.4 |
| Sugar cane, Hawaii | 19 | 380 | 1.6 |
| Sugar cane, Java | 20 | 560 | 2.2 |
| Broadleaf forests, Southeast U.S. | 19 | 100 | 0.40 |
| Temperate zone bogs | 17 | 160 | 0.70 |
| Subtropical bogs | 21 | 840 | 3.3 |
| Tropical bogs | 19 | 970 | 4.0 |
| Marine systems |  |  |  |
| Sargasso Sea | 20 | 22 | 0.08 |
| Laminaria Community, Nova Scotia | 14 | 280 | 1.6 |
| Peru Current | 16 | 590 | 2.8 |
| Coral Reef | 23 | 680 | 2.2 |

Poole and Williams

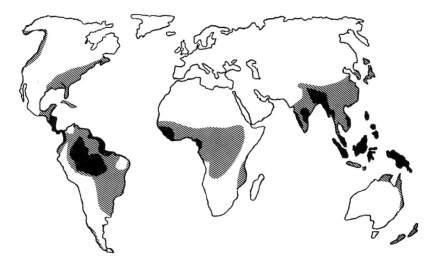

**Figure 9.3**
Pattern of world precipitation. Most high precipitation areas lie in the tropics. While high precipitation alone is not an index of high potential biomass yields, it is an important factor. White areas indicate less than 40 inches of annual precipitation, shaded areas indicate 40 to 80 inches of annual precipitation; and black areas indicate greater than 80 inches of annual precipitation.

Nutrient availability can limit photosynthesis in some ecosystems. This is particularly true in marine systems, as illustrated by some of the examples in table 9.1. In the subtropical Sargasso Sea, where there is nearly 50 percent more sunlight than the global average, yields are actually less than the global average. This arises because the sun heats surface waters, resulting in higher density water at lower depths, thereby suppressing convective mixing of deep and surface waters. Such mixing is necessary for high productivity since deep ocean waters tend to be nutrient rich and surface water nutrient poor. In contrast to the Sargasso Sea, the Peru current is highly productive, owing to the fact that nutrients are brought to the surface through natural upwelling. From the figures cited in table 9.1 for Nova Scotia, it can be seen that production is higher in northern surface waters: the surface layer is not especially warm, so nutrients tend to be more available.

It is interesting to note in table 9.1 that agriculture is not especially productive relative to many natural ecosystems. One reason for this is that many crops are annuals so that each year during the early part of the growing season available sunshine is inefficiently utilized before a full crop cover is established. And unlike agriculture, which is dominated by single species ecosystems (or monocultures), natural ecosystems typically have a number of plant species which may have complementary as well as competitive strategies for utilizing resources such as light and water. Also crops are generally raised to maximize economic value for food or fiber rather than to maximize the total photosynthetic yield.

### Converting Plant Biomass

Plant biomass can be used as fuel in a number of different forms. For a given plant biomass source the choice of a suitable fuel form depends mainly on the characteristics of the biomass material, the market values of alternative fuel forms, and the location of markets relative to the biomass source. The last factor here involves consideration of transport costs of alternative fuel forms.

In our analysis we have found it useful to classify conversion processes according to the quality of fuel produced, as shown in figure 9.4. In one conversion stream a "high quality" fuel functionally analogous to the higher fractions of petroleum or natural gas is produced. In the other a "low quality" boiler fuel, analogous to coal or residual fuel oil, is produced. For both the high and low quality fuel forms plant biomass offers potentially significant advantages over alternative fuel inputs.

The higher quality fuel stream is particularly attractive because the major conventional fuel options for the future (coal and nuclear power) are best suited for electric power generation and not for producing the high quality liquid and gaseous fuels that are better suited for providing certain demands. There remain significant technological obstacles to the electrification of auto or truck transportation, to the production of hydrogen, and to the production of methane and high quality liquid fuels from coal at competitive costs.

While it is premature to make firm prognostications until more experimental work is done, it appears that a fermentation of nonwoody biomass, yielding methane, may be the most economical conversion technology for producing high quality fuels from some forms of plant biomass. This biogasification process will likely yield methane at less cost than hydrogen from water via electrolysis [5] or thermochemical processes.[6] The conversion costs for producing methane this way may well be competitive with those for producing methane from coal.[7] Further, methane produced

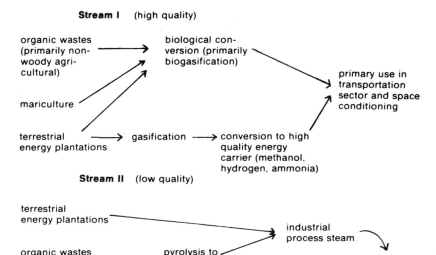

Figure 9.4
Simplified scheme of major energy streams.

in biogasification is likely to be considerably less costly than methanol produced via pyrolysis from plant biomass.[8]

An intrinsic advantage of the biogasification process is that it is relatively simple from an engineering point of view, compared to alternative processes for producing high quality fuels. This factor and the likely future competitiveness of the process, even at relatively small scale operations, suggest the possibility of widespread application of this concept.

Cost is not the only consideration, however, in selecting a suitable fuel form. Methanol, a major alternative high quality fuel, is inherently more valuable than methane because it is a liquid and thus more suited, for example, as a transport fuel. Methanol has been praised recently as an ideal fuel to replace gasoline.[8] Indeed our transportation system would have to be modified little to accommodate a shift to methanol.

Pyrolysis is an important conversion process for woody biomass which is not suitable for conversion to methane via biogasification.[9,10]

In pyrolysis organic material is heated in an oxygen deficient atmosphere to convert biomass to a form (or forms) more suitable than raw biomass for some applications. Depending on the conditions of the conversion process the output could be predominantly a solid fuel (char), a tarry

Flower Power

liquid fuel, methanol, a gaseous fuel (primarily carbon monoxide and hydrogen), or a mix of components.

Pyrolysis yielding a char or crude tarry liquid is a relatively simple, low cost, high efficiency conversion process,[10] which is well suited for boiler fuel applications where the biomass must be shipped long distances to industrial centers. The products of pyrolysis would be cheaper to transport than raw biomass, having more energy per ton and being easier to handle and store.

Biogasification and industrial boiler fuel applications will now be discussed in somewhat more detail.[11]

*Biomass Fermentation to Methane* Biogasification will convert most nonwoody biological materials found in abundance in nature to methane and carbon dioxide. The product gas, when scrubbed of carbon dioxide and small quantities of other gases, is essentially the same as pipeline quality natural gas. Materials rich in lignin, such as wood, are not suitable because lignin is resistant to fermentation and its presence may limit or altogether block the digestion of some plant materials. But fresh nonwoody plant biomass and manure with a high moisture content may be more suited for anaerobic fermentation than for alternative biomass conversion schemes.

While the state of the technology for anaerobic fermentation is not yet fully developed, it is not unrealistic to expect an energy conversion efficiency of 50 percent and a conversion process cost in the range of $1.25 to $2.00 per million Btu of output for conversion facilities processing 100 tons of dry matter or more per day.[12]

Below 100 tons per day, conversion costs rise sharply, according to available evidence.[13] A minimum scale of 100 tons per day, however, should not be a serious constraint in tapping biomass resources, including relatively diffuse crop residues, as we discuss below.

A characteristic feature of biogasification is that an indigestible residuum remains. This residuum is an excellent nutrient source and soil conditioner for recycling onto agricultural land or back onto energy "plantations." In some cases it may even be a respectable high protein feed for ruminants.

The overall cost of methane from biogasification involves both the process cost and the cost of biomass raw material. As we show below, the biomass cost may range from about $0.30 per million Btu for feedlot manure or crop residues to $1.00 for energy plantations. Thus it is reasonable to expect that the overall cost of methane produced this way could lie between $2 and $4 per million Btu. For comparison, methane from coal is expected to cost between $2.50 and $4.00;[7] hydrogen produced via electrolysis on the order of $6.50;[5] and hydrogen produced via thermochemical processes, if such a technology can be successfully developed, $3.50 or more[6] per million Btu. It is reasonable to conclude that methane produced via biogasification could become one of the most promising synthetic fuels.

*Plant Biomass as Industrial Boiler Fuel* Some work has been done on the utilization of low quality fuels from biomass as boiler fuel at central station electric power plants.[14,15] It is our view that this application is far less desirable than use as industrial boiler fuel.

One reason for this judgment is that while there are many new energy sources that will be suitable for central station power generation, over the long term biomass appears to be the only major alternative to coal and other fossil fuels for industrial process steam use.

Another reason involves consideration of the appropriate scale for biomass operations. In general, unit costs will vary with the size of any technological operation. For small scale operations unit costs are high. As the scale of operations increases, unit costs decrease, that is, scale economies are achieved. In most cases, however, a point of diminishing returns is eventually reached, so that beyond some critical size costs will tend to increase again with greater scale.

In the case of central station electricity generation based on coal or nuclear fuels significant scale economies are achieved with plant sizes up to 600-700-megawatts (electrical), and in fact most new plants are being built to capture even further scale economies with sizes up to 1,000- or 1,200-megawatts (electrical). For a biomass electric power plant to compete with say, a coal-fired plant, the total annualized costs—mainly capital charges plus delivered fuel costs—must be comparable to those for the coal plant.

The critical scale of operations for a biomass plant (where costs are minimized) is determined by the scale economy curves for both the conversion plant (which would be similar to that for a coal-fired plant) and the fuel collection and delivery system. Unfortunately, the cost of delivered fuel is likely to rise very steeply before significant scale economies can be realized in the size of the conversion plant. The reason for this is that a prohibitively large land area must provide the fuel for a 1,000-megawatt (electrical) baseload power plant. Such a power plant (operating on the average at 60 percent of capacity) requires biomass fuel delivered to one site at a rate of about 10,000 tons of dry matter per day. A plantation of nearly 400,000 acres would be required to fuel a 1,000-megawatt (electrical) plant, assuming a yield of 10 tons of dry biomass per acre per year.

As discussed below conditions for biomass plantations appear to be most favorable in the southeastern United States. But, as pointed out in a recent report,[14] land in large blocks is probably not available east of the Mississippi to support plants with this scale of operations.

The appropriate scale of operations for a biomass production facility depends on many factors, but it is likely to be somewhere in the range of 100 to 1,000 tons per day of dry matter instead of 10,000. The critical scale for biomass production and collection appears to be compatible with the scale of industrial process steam operations. More than half the process steam produced in this country is at installations using 400,000 pounds of steam per hour or less,[16] which is equivalent to biomass fuel use of 800 tons of dry biomass per day or less.

Burning fuel to produce only process steam, in the case of biomass or fossil fuel, involves a gross waste of fuel resources. It would be more economical to first generate electricity with the energy released in combustion and then use the waste heat for process steam applications. In this "cogeneration" of steam and by-product electricity, the amount of fuel required to produce electricity, beyond what is required for process steam alone, amounts to only about half of what is required at a central station power plant. In the new era of high cost fuels, cogeneration is economical, even for quite small industrial power plants using various fossil fuels.[16,17] In most cases economics also favor use of biomass for electricity production via industrial cogeneration over electricity production at a central station power plant.

There are two different biomass cogeneration configurations of interest. In one the steam-using industry would be located at the biomass plantation, in a true "industrial park." In this case the dry biomass could be used directly as fuel. If the biomass is to provide a significant fraction of the total industrial steam load, however, it would be undoubtedly necessary to transport some of the biomass in a suitable form (such as pyrolytic char) considerable distances to steam-using industries that can't for one reason or another be sited at the plantations.

## Large Scale Production

Producing biomass in quantities sufficient to meet a significant fraction of U.S. energy demand will require an intensive use of resource inputs. In this section we survey some of the principal resource problems that may limit the output of terrestrial and marine biomass production systems and explore strategies for minimizing these problems. Water and land availability are key limiting factors for terrestrial production, while difficulties associated with nutrient supply may be a serious problem for mariculture. Detailed descriptions, of different biomass plantations can be found in the literature.[14,15,18,19]

*Terrestrial Systems* Water resource availability is likely to be one of the most significant factors limiting increased photosynthetic production on land. On the order of 500 tons of water are needed typically for each ton of dry plant matter produced.[20]

The water resource constraint is reflected in the potential average yield one could expect from a large terrestrial solar plantation. Consider a favorable area such as the southeast United States where rainfall averages some 48 inches per year. If all the runoff from the plantation were retained for irrigation, this rainfall would correspond in a natural ecosystem to a yield of about 10 tons per acre per year (or a photosynthetic conversion efficiency of about 0.65 percent). Irrigation sources from outside the plantation would be required to increase yields further.

The external irrigation water requirements to double this yield, for example, and provide 5 quadrillion Btu of useful energy for the nation as a whole (some 35 million acres committed to the solar plantation), would amount to more than half the total freshwater runoff in the South Atlantic-Gulf water basin outside the plantation. The example illustrates the importance of designing and locating large plantations so as not to depend on substantial external irrigation. Thus areas in the southwest United States would be inappropriate for producing substantial quantities of biomass.[21] The example also suggests that when selecting the appropriate plant species for a biomass plantation it may be more important to emphasize low transpiration water requirements than high potential yields. Some species require much less water for transpiration than others.[20]

Because solar energy is diffuse, land is another major resource constraint for the terrestrial solar plantation. High photosynthetic yields are often achieved on land that has high value for other purposes. One strategy for minimizing the land constraint is to simultaneously use the land for some other productive purpose such as agriculture or forestry and produce fuel materials as a by-product from residues, as described below. If, however, photosynthetic energy is to be developed beyond the somewhat limited potential of by-product energy resources, attempts should be made to use land that is reasonably productive of biomass but has relatively low value for alternative uses such as forestry, preservation of wilderness, recreation and, especially, agriculture.

Given the importance of water for biomass production, one promising class of land is that which is too wet for agricultural use. Indeed very wet lands, such as swamps and marshes, often carry the most productive ecosystems. Sustained harvested biomass yields of about 10 tons per acre per year should be possible. We estimate that 100 to 150 million acres of productive land with wetness problems and low potential for agricultural use exists in the United States and might be available for biomass production[22] (see figure 9.5). The conflict with forestry products may be more significant. Present forest management strategies are probably incompatible with such an expansion of acreage devoted to biomass production. But there appear to be opportunities to substantially increase production of conventional forest products.[23]

The great bulk of this land does not appear to be synonomous with wetlands which are regarded as prime wilderness areas, but is rather land characterized by periodic water logging. Wetlands that are notable as wilderness areas and rich habitats for wildlife should be preserved. It should be noted, however, that in some cases management of wetland for biomass production may have a far smaller impact on the ecosystem than conventional agricultural development, judging from relevant experience in England[24] and the Danube delta in Rumania.[25] Nevertheless analysis of potential impacts should be carried out in evaluating the biomass production potential of large wetland resources.

Besides land with wetness problems other types of land may be suited for biomass plantations. Some other forest lands may be quite productive and still have low enough opportunity costs to justify biomass production.[26]

Some form of management of the ecosystem will generally increase harvested yield, but the type of management which is appropriate is not yet clear and is in any case likely to vary from area to area. Highly intensive management closely resembling current agricultural practice with crops such as sorghum or sugar cane has been favored in most discussions. It appears that with this approach production costs of $1.00 per million Btu of biomass are achievable.[14] Unfortunately, this strategy is based on the assumption of the availability of large land areas which are highly suited to large scale production of agricultural crops. Many field operations associ-

Poole and Williams

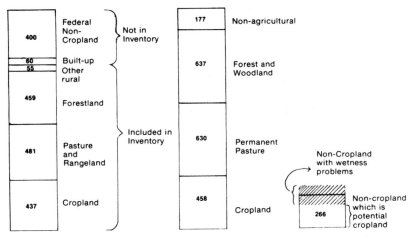

Source: National Inventory
of Soil and Water Conser-
vation Needs, 1967, U.S. Department of Agriculture,
Statistical Bulletin No. 461.

Source: The Nation's
Water Resources,
U.S. Water Resources
Council, Washing-
ton, D.C., 1968.

Source: USDA Statis-
tical Bulletin No. 461.

**Figure 9.5**
Land-use in coterminous United States. The two bar graphs on the left represent alternative breakdowns, in millions of acres, of current U.S. land use. The bar graph
on the right illustrates (in the same scale) the area of non-cropland which is potential
cropland according to the U.S. Department of Agriculture. Overlaid on this is the
area of non-cropland which is limited for agricultural use by wetness problems; the
area of overlap represents the area of potential cropland which is limited by wetness
problems.

Flower Power

ated with conventional agriculture are costly or impossible on land with low opportunity costs. Thus an appropriate strategy for using such land may require novel and perhaps rather simpler manipulations of the ecosystems (such as short rotations of hardwood trees[15,26] or the reed culture of the Danube delta [25]).

It is frequently assumed that biomass production would be a very large-scale affair. Blocks of 50,000 to 150,000 acres of land [14,15] are the usual scale quoted—hence the term plantation. However, it is doubtful that operations producing biomass would really be subject to scale economies beyond a relatively modest size of, say, 1,000 to 2,000 acres. Conventional crop production, which probably requires more separate specialized pieces of equipment than biomass production, involves no significant scale economies beyond 1,000 acres.[27] This suggests that we might pay more attention to the possibility of organizing biomass production around "farms" rather than large plantations. The relationship of these farms to conversion facilities might well resemble that of, say, sugar beet producers to sugar mills. This institutional approach may well reduce siting problems, since the problems resulting from attempts to assemble large acreages under single ownership or management would be avoided.

It is possible that within the collection area of a conversion facility agricultural crops could be grown on prime cropland and biomass on the remaining land, depending on the opportunity cost of such land. Agricultural residues could then flow into the conversion facility as well. Since relatively small-scale conversion technologies exist, this approach may well be viable. It would permit more selective and appropriate use of different land types in a region and should consequently increase the resource base available for biomass production without using land more suitable for other purposes.

Nutrient supply may be a long term concern. Work at the Institute for Energy Analysis suggests that while most essential minerals can be supplied perennially at fairly nominal energy cost in a worldwide steady state society with average per capita material consumption comparable to the United States today,[28] there may be several exceptions. Prominent among these are phosphorus[29] and certain trace minerals of importance to agriculture. The long term impact of these possible nutrient constraints requires careful study. Will recycle of nutrients be an adequate response? If so, what technical and institutional constraints may limit recycle?

*Marine Systems* Ocean farming systems offer the possibility of enormous expansion of biomass production, with the principal constraints of terrestrial systems (available space and water) largely eliminated. Interest in mariculture is now focused on a floating kelp farm concept.[30] The farm consists of a grid of lines to which the kelp are attached. The grid would serve to maintain proper spacing of the kelp and would be adapted to prevent storm damage to the kelp. Yields of 30 tons per acre are believed to be realistically achievable at lower latitudes, if the kelp are kept in an environment of cold nutrient-rich water. Kelp has been chosen as the experimental organism because it has a relatively high growth rate and methods for propagating and harvesting it are well understood. After harvesting most kelp processing would take place on land initially. The kelp will be tested both as an animal feed and as feedstock for biogasification.

Mariculture technology for biomass production is embryonic. Thus it is premature to estimate costs for commercial operation. Nevertheless it is clear that mariculture will be much more capital intensive than terrestrial biomass production. The grid structure for attaching the kelp will account for much of the required capital.

Yields of biomass in the open ocean are generally low, owing mainly to the low concentration of nutrients in surface waters, where photosynthesis occurs. With a high capital investment per unit area, high yields obtained from increased nutrient concentrations are essential to this concept.

One direct approach to obtaining more nutrients is to fertilize the surface layer.[18] This approach, however, involves a serious constraint that could limit the scale of biomass production. A key nutrient is phosphorus, which makes up about 0.4 percent of the dry weight of biomass such as kelp. In principle the phosphorus taken up by the kelp can be recycled (for example, recovered from the residuum of a digester if the kelp is processed via biogasification). But some of the applied phosphorous will be lost via gravitational settling, convection, and diffusion from the kelp field. Experimental evidence is inadequate to ascertain what fraction of the applied phosphorus is lost this way. But one estimate is that the loss amounts to 50 percent.[18]

At this rate methane production from kelp at a level of 5 quadrillion Btu per year would result in a direct phosphorus loss amounting to 2.5 million tons a year, which is about 18 percent of total world phosphorus consumption. This level of phosphorus consumption is a serious concern because of the potential for depleting high grade phosphorus resources. The amount of such resources is large, adequate perhaps for a couple thousand years at the present rate of world consumption.[29] But phosphorus is almost unique among mineral resources in that it is essential for life itself and other substances cannot be substituted for it as an essential nutrient.

Until a better understanding of world phosphorus resources is acquired and appropriate phosphorus conservation policies are established it would be imprudent to allow a large increase in world phosphorus consumption for such a small increase in energy supplies, as this mariculture nutrient calculation suggests. We would emphasize again that this calculation is based on the assumption that 50 percent of the applied phosphorus is lost. It may be possible to design systems where the loss can be kept to very low levels.

An alternative strategy for supplying nutrients to surface waters is to create an artificial upwelling, bringing nutrient-rich deep ocean water to the surface. This upwelling of cold water is also desirable because at lower latitudes surface waters are too warm for optimal growth.

Unfortunately bringing deep ocean water to the surface results in a release of carbon dioxide to the atmosphere, owing to the fact that deep ocean waters are enriched in inorganic carbon relative to surface waters. In fact, a simple calculation[31] indicates that about three times as much carbon dioxide is released per Btu of methane produced from mariculture biomass than is released from the burning of a Btu of fossil fuel methane. This is a disturbing potential side effect of upwelling mariculture, especially in light of the fact that a major motivation for using photosynthetic energy is to avoid the possible climatic problems associated with the atmospheric build-up of carbon dioxide. Unless a technique can be developed to avoid this problem energy from upwelling mariculture may not be environmentally preferable to the continued burning of fossil fuels.

## Organic Waste Resources

As we observed earlier, the use of by-product crop residues (such as corn stalks) is one way to relax the land constraints on biomass production. These and other organic wastes represent a considerable energy resource, amounting to about 11 to 12 quadrillion Btu per year, as shown in table 9.2. (For comparison, coal consumption in the United States is currently about 13 quadrillion Btu.) Thus, this basic resource, organic wastes, ranks in the energy major leagues, though, of course, we must expect losses in collection and conversion to more useful energy forms.

Once it has been realized that a substantial energy resource exists in wastes, the key problem will be to collect wastes together on a scale large

Poole and Williams

**Table 9.2**
Organic waste resources in United States (1970-1975)

| | Btu per year (quadrillions) |
|---|---|
| Crop residues[a,b] | 5.2 |
| Waste wood in forests[a,c] | 4.5 |
| Manure from animals in confinement[a] | 0.7 |
| Urban refuse[b] | 1.3 |
| Total | 11.7 |

[a] Assuming an energy content of 15 million Btu per ton of dry organic matter.
[b] See Poole.[12]
[c] There is considerable disagreement on the magnitude of this resource, due in part to differences in the definition of the waste resource. Current estimates range from about 500 million tons dry matter cited in Tatom[10] to about 150 million tons (report of work at Stanford Research Institute given by John Alich at the Conference on Capturing the Sun Through Bioconversion, March 10-12, 1976). We have assumed approximately 300 million tons.

enough to make conversion to useful energy forms economically interesting. The appropriate scale of operations depends on the technology, as we discussed earlier; but with 50 to 100 tons per day at one site, scale-cost relationships of some conversion technologies begin to look favorable. We shall consider a minimum scale of 100 tons per day dry matter when assessing the difficulties in collection.

*Crop Residues* How far crop residues have to be transported to supply a conversion facility with 100 tons of residue per day varies widely from region to region. It depends sensitively both on the fraction of the land area in fairly small blocs (say, 10 miles square) planted in residue yielding crops and the residue yield per acre. We estimate that perhaps 40 percent of the land area in the major crop producing areas of the country would yield residues and that a typical residue yield would be on the order of 2 tons per acre. For these conditions the area of the wasteshed required to supply a 100 ton per day plant is about 70 square miles and the average haul is less than 3.5 miles, if the plant is at the center of this area. The energy costs of collecting and transporting material this far are trivial, about 0.1 million Btu per ton, and farmers should be able to deliver residues directly to the plant from the fields for 25 to 30 cents per million Btu of biomass.[12] A conversion technology such as biogasification may well be favored for crop residues so that the nutrient bearing organic residuum could be returned to the soil. While returning the sludge to the soil may add to the cost of the system, the extra cost is likely to be quite small.[12]

*Animal Manure* Animal manure has attracted considerable interest as a feedstock for energy conversion. But so far most of this interest has been on biogasification as a means of ameliorating a serious waste disposal problem at particular sites where large quantities of animal manure are concentrated. With the exception of a few large feedlots, the scale of these operations has usually been such that fuel production is distinctly uneconomic without large credits for disposal costs saved.

A more promising approach to energy conversion for manure is to collect manure from a number of enterprises, just as we have proposed for crop residues. Gathering the wastes is facilitated with this strategy by the regional specialization for the major animal stocks likely to provide the bulk of manure in confined conditions. Also the considerable overlap between many livestock and crop production regions reduces the area of the "wasteshed" required for a conversion facility of economic scale below what it would be for crops or livestock considered separately.

*Forest Wastes* Forest wastes are available in essentially two forms. Annual litterfall and scattered mortality of trees constitute a diffuse resource, while logging, culling, and mass tree mortality[32] generate more concentrated wastes. The numbers available on these resources are "soft," but a resonable estimate is that somewhat more than half the forest residues fall in the second category. This is fortunate since the concentrated residues are more accessible, and their exploitation probably raises fewer environmental questions. The concentrated sources may yield 10 tons of residue or more per acre. A promising technology for the exploitation of these forest wastes is a mobile pyroltic plant yielding char-oil.[10]

### Potential Impact

The extent to which photosynthetic energy will eventually be developed depends on the costs of these fuel forms relative to alternatives. While a detailed assessment of costs is beyond the scope of this report, our preliminary survey of costs indicates that once development of photosynthetic energy is underway the costs would tend to be greater than those of conventional fuels but are likely to be competitive with many of the better studied alternative substitutes for conventional fuels. However, beyond some level of development, marginal costs of photosynthetic energy production would rise sharply as limits on resource inputs (available land, water, etc.) are approached.

The level of photosynthetic energy production at which these marginal costs will be limiting is uncertain at this time. But even if photosynthetic development is restricted to areas and levels where limiting factors do not appear to be important, this energy source can meet a significant fraction of total energy demand.

To illustrate the magnitude of this resource potential for the United States, a hypothetical photosynthetic energy supply budget is shown in table 9.3. For the sake of simplicity, we consider only two conversion routes: biogasification of nonwoody organic materials and cogeneration of steam and electricity for other materials.

Our estimate is that about a fifth of today's energy needs could be supplied by photosynthetic energy before limiting factors become important. The assumptions underlying this estimate are summarized in the notes to table 9.3. In the case of the terrestrial plantations the amount of land considered (125 million acres) appears to be within the range of potentially available productive land with relatively low opportunity costs. Similarly the assumed yields for these plantations (10 tons per acre for nonwoody and 6 tons per acre for woody dry biomass) appear to be within reach of current technologies for such land. Indeed, very much higher yields may be obtained in some parts of the country.[33]

Perhaps the most optimistic assumption in the budget is that 80 percent of the more accessible wastes (manure available from confined animals, field crop residues, concentrated forestry wastes, etc.) could be collected. We have chosen a high recovery rate to highlight the fact that with a concerted effort aimed at waste recovery the benefits are substantial—some 5 quadrillion Btu of usable energy per year.

Research and engineering development will be necessary to realize these goals, and there is the possibility that target costs cannot be reached. But the development and cost problems are probably less severe than with many other new energy technologies, and the success of photosynthetic fuel systems is not dependent on the success of any single technology. A variety of options exist for biomass supply and conversion.

One interesting possibility for stretching the energy potential of biomass is to convert it to hydrogen using a solar furnace, as described in Antal's essay in this volume. For a given amount of biomass the potential

**Table 9.3**
Biomass energy supply

| | Energy (quadrillion Btu per year) |
|---|---|
| Methane from biogasification | |
| Crop residues[a] | 2.0 |
| Manure[b] | 0.3 |
| Plantation (nonwoody)[c,d] | 5.6 |
| Total | 7.9 |
| Low quality fuels | |
| Urban waste[e] | 0.8 |
| Logging waste[f] | 1.0 |
| Pulp waste[g] | 0.6 |
| Plantation (woody)[h] | 3.6 |
| Total | 6.0 |
| Equivalent value of low quality fuels when used for industrial cogeneration[i] | 7.7 |

[a] Assume that 80 percent of 340 million tons is collectable and that biogasification is 50 percent efficient in converting organic materials to methane.
[b] Assume that 80 percent of 45 million tons produced by animals in confinement is collectable and converted to methane at about 50 percent efficiency.
[c] For 75 million acres yielding 10 tons of biomass per acre.
[d] Mariculture has not been included in the budget because both the microeconomics and resource constraints are particularly speculative at this time. The system could augment the resources here very considerably. This would be true even if mariculture proved to be too expensive for fuel production, but was a competitive source of fodder, since any factor reducing the land requirements of agriculture and forestry would permit a greater allocation of land to terrestrial fuel plantations.
[e] Char oil produced at 70 percent efficiency from collectable urban refuse generated at the 1973 level.
[f] Char oil produced at 70 percent efficiency from 100 million tons of logging wastes and other concentrated forest and sawmill residues.
[g] Use directly 40 million tons (1973 level) of pulp waste as a coal substitute.
[h] Assume 50 million acres at about 6 tons per acre. Assume one-half of energy is from biomass used directly as a coal substitute and the other half is from char oil produced at 70 percent efficiency.
[i] This assumes that on the average about 100 kilowatt hours of electricity is produced per million Btu of steam generated, and that the extra fuel required to produce electricity is about half that required at a central station power plant. Thus about 2.2 quadrillion Btu per year would be saved compared to generating this electricity at a central station power plant.

quantity of hydrogen produced could amount to more than three times the amount of useful energy generated via biogasification.

In this report we have focused almost wholly on the prospects for using biomass as an energy source for the United States. But as we have already pointed out, and as figure 9.3 shows, the parts of the world where the greatest potential exists for utilizing biomass as fuel are outside the United States. In fact, some of the most energy starved countries in the world are in regions where the greatest potential exists for this energy source.[4] Also it appears possible to develop biomass technologies keeping capital inputs low in relation to labor inputs.[34] This is particularly important for developing nations where capital is scarce and costly but labor is abundant and cheap. Moreover, as we have pointed out, the scale of economic operations for biomass conversion can be much smaller than for, say, a nuclear power plant, so that there is much greater flexibility In introducing new capacity to match energy needs. Clearly, photosynthetic energy for developing nations should be a high priority item for research and development.

## Notes

1. For a general discussion of the climatic implications of the atmospheric build-up of carbon dioxide, see *Inadvertent Climate Modification: Report of the Study of Man's Impact on Climate (SMIC)* (Cambridge, Mass.: MIT Press, 1971). More recently evidence has been presented that the present global cooling trend may be one phase of a long-term climate cycle which can be expected to end soon. Wallace S. Broecker has argued that we may be on the verge of a pronounced global warming from the increased carbon dioxide in the atmosphere. See *Science* (Aug. 8, 1975), p. 460.

2. L. E. Rodin and others, "Productivity of the World's Main Ecosystems," in *Productivity of World Ecosystems*, Proceedings of an NAS Symposium, Seattle, Washington, Aug. 31–Sept. 1, 1972 (Washington, D.C.: National Academy of Sciences, 1975).

3. Pure glucose has a heat value of 3.7 kilocalories per gram. From R. S. Loomis and others ("Agriculture Productivity" in *Annual Review of Plant Physiology*, 22 (1971), 431), we obtained energy contents of 4.6 to 4.9 kilocalories per gram for woody plants and 3.4 to 4.4 kilocalories per gram for others.

4. In these regions which make up about one percent of the continental land area of the world, a recent estimate [2] is that some 33 billion tons of biomass or about 500 quadrillion Btu of energy are produced in photosynthesis each year—more than twice the present level of world energy consumption. Production in these regions (in billions of tons per year) is distributed as follows:

| | |
|---|---|
| Subtropical bogs | 4 |
| Subtropical floodplains | 3 |
| Tropical swampy forest | 6 |
| Tropical bogs | 11 |
| Tropical floodplains* | 9 |
| Total | 33 |

Flower Power

*Excludes 9 billion tons per year on arid or semiarid floodplains where agricultural activity is concentrated.

5. Electrolysis plants currently operate at an efficiency of about 60 percent. Advanced systems may achieve 80 percent efficiency. With this improvement the cost of hydrogen per million Btu would be 0.312 × (cost of electricity in mills per kilowatt hour) + 0.227.

With electricity from new baseload power plants costing 20 mills per kilowatt hour, electrolytic hydrogen would cost $6.50 per million Btu. For details see R. H. Wentorf and R. E. Hanneman, "Thermochemical Hydrogen Generation," *Science* (July 26, 1974), p. 3.11.

6. This is based on the use of a nuclear plant producing heat only at $1.50 per kilowatt (thermal), a 90 percent capacity factor, a nuclear fuel cycle cost of $0.20 per million Btu, a 50 percent conversion efficiency for making hydrogen, and hydrogen processing costs of $1.00 per million Btu. See Wentorf and Hanneman [5].

7. D. C. White and O. H. Hammond, "Research in Coal Combustion Appears the Cinderella of U.S. Energy Effort," *Energy Research Reports* 2:2 (Feb. 1976).

8. See T. B. Reed and R. M. Lerner, "Methanol: A Versatile Fuel for Immediate Use," *Science* (Dec. 28, 1973); T. B. Reed, "Biomass Energy Refineries for Production of Fuel and Fertilizer," paper presented at Eighth Cellulose Conference, May 20–22, 1975, Syracuse, New York.

9. There is considerable development of pyrolysis technologies today. See, for example, "Feasibility Study Conversion of Solid Waste to Methanol or Ammonia," report to the City of Seattle Department of Lighting, by Mathematical Sciences Northwest, Sept. 6, 1974; and "Fuels from Municipal Refuse for Utilities: Technology Assessment," report prepared for the Electric Power Research Institute by Bechtel Corporation, March 1975. For reasons of space we have not discussed pyrolysis in detail.

10. J. W. Tatom and others, "A Mobile Pyrolytic System for Conversion of Agricultural and Forestry Wastes into Clean Fuels," 1975 preprint from Engineering Experiment Station, Georgia Tech, Atlanta, Georgia 30332.

11. The conversion options considered here do not exhaust the possibilities. Breakthroughs, for example, with the cellulose-ethanol conversion scheme or with hydrogen-yielding schemes may make such alternatives attractive for commercial energy applications. (See, for example, *Proceedings of the Workshop on Bio-Solar Conversion*, a report on an NSF sponsored workshop held Sept. 5–6, 1973 at Bethesda, Maryland.) In addition, there are potential applications as a petrochemical feedstock. See I.S. Goldstein, "Potential for Converting Wood into Plastics," *Science* (Sept. 12, 1975), pp. 847–852.

12. Alan Poole, "The Potential for Energy Recovery from Organic Wastes," in *The Energy Conservation Papers*, ed. Robert H. Williams (Cambridge, Mass.: Ballinger Press, 1975).

13. C. N. Ifeadi and J. B. Brown, Jr., "An Assessment of Technologies Suitable for the Recovery of Energy from Livestock Manure," Battelle Columbus Laboratories report, June, 1975; and G. R. Morris, W. J. Jewell, and G. L. Casler, "Alternative Animal Wastes Anaerobic Fermentation Designs and Their Costs," preprint from Cornell University, 1975.

14. J. A. Alich, Jr., and R. E. Inman, "Effective Utilization of Solar Energy to Pro-

duce Clean Fuel," Stanford Research Institute report to the National Science Foundation, June 1974.

15. G. C. Szego and C. C. Kemp, "Energy Forests and Fuel Plantations," *Chemtech*, May 1973.

16. "Energy Industrial Center Study," report prepared for the National Science Foundation, by the Dow Chemical Company, the Environmental Research Institute of Michigan, Townsend-Greenspan and Company, and Cravath, Swaine, and Moore, June, 1975.

17. R. H. Williams, "The Potential for Electricity Generation as a By-Product of Industrial Steam Production in New Jersey," Center for Environmental Studies Report, Princeton University, May, 1976.

18. E. J. Szetela and others, "Technology for Conversion of Solar Energy to Fuel," report prepared for the University of Pennsylvania by United Aircraft Research Laboratories, March 1974.

19. Bill McLean and Leroy Riggs, "Progress Report on Open-Ocean Marine Energy Farm Project," Naval Undersea Center at San Diego, Oct. 18, 1973.

20. From 300 to 750 tons of water (depending on species) are required for transpiration alone to produce a ton of dry plant matter. See C. C. Black, "Ecological Implications of Dividing Plants into Groups with Distinct Photosynthetic Capacities," *Advances in Ecological Research*, vol. 7 (1971).

21. In the case of a recent proposal to produce biomass in the dry Southwest [14], the water requirements for producing 5 quadrillion Btu of useful energy would be even more astronomical. In this proposal a yield of 30 tons per acre is considered. With minimal transpiration water requirements of 300 tons per ton of dry matter plus delivery losses of 2 acre-feet per acre, total water requirements would be 170 billion gallons per day or 40 percent of all the available runoff within a thousand miles.

22. Based on data in the 1967 *Basic Statistics-National Inventory of Soil Water Conservation Needs* (Soil Conservation Service, U.S. Department of Agriculture, Statistical Bulletin No. 461, January 1971). The inventory classified all non-federal cropland and non-cropland by the primary factor limiting use as cropland. The four factors considered were: erosion, wetness, climate and soil deficiencies (for example, too stony or saline). The intensity of the limitation is ranked on a scale of increasing severity from I to VIII. As shown in figure 9.5, non-cropland where wetness is the primary limiting factor is estimated to be 164 million acres, and approximately half (76 million acres) of this cropland is considered as potential cropland (limitation of class IV or less). However, it appears that only about a third of this 76 million acres has a significant potential for agricultural applications. (See R. C. Otte and H. T. Frey, "Potential Cropland," inhouse study of Economic Research Service of U.S. Department of Agriculture, Aug. 6, 1974; and M. L. Cotner et al., "Land Resource Capabilities for U.S. Food Production," unpublished paper presented at annual meeting of American Association for Advancement of Science, New York, Jan. 1975).

23. About 70 percent of the 164 million acres of non-cropland with wetness problems is also classified as forest land. This is about 23 percent of the commercial forest land in the coterminous United States. Whether anything like this fraction of commercial forest land can be removed from conventional fiber production may largely depend on the success of measures to intensify fiber output on a significant fraction of forest

land. For a discussion of some development work, see C. L. Brown, "Forests as Energy Sources in the Year 2000," *Journal of Forestry*, 74:1 (January 1976).

24. The reedswamp at Wicken Fen, near Cambridge, England is regularly harvested for thatching and is also maintained as a bird sanctuary by the Nature Conservancy. See articles by S. M. Haslam and D. S. A. McDougall, in *The Reed ("Norfolk Reed"),* Norfolk Reedgrowers Association, 1972. Obtainable from S. M. Haslam, at the School of Botany, Cambridge University, Cambridge, England.

25. L. Rudescu, C. Niculescu, I. P. Chiou, *Monografia Stufului Din Delta Dunari,* Academiei Republicii Socialiste Romania, 1976 (Buscuresti, Colea Victoriei No. 27).

26. Four to five tons dry weight of fiber is the average annual yield achieved over a variety of sites in Georgia with the short rotation hardwoods developed at the School of Forest Resources, University of Georgia. Biomass yields are higher. Fertilization is minimal and genetic improvements are anticipated. See K. Steinbeck, personal communication to Alan Poole, February 1976, and K. Steinbeck and C. L. Brown, "Yield and Utilization of Hardwood Fiber Grown on Short Rotation," *Applied Polymer Symposium No. 28* (New York: John Wiley and Sons, Inc., 1975).

27. Walter W. Wilcox et al., *Economics of American Agriculture* (3rd ed.; Englewood Cliffs, N.J.: Prentice-Hall, 1974).

28. H. E. Goeller and Alvin Weinberg, "The Age of Substitutability," *Science* (Feb. 20, 1976), p. 683.

29. Workshop on Global Ecological Problems, *Man in the Living Environment,* Institute of Ecology, Madison, Wisconsin, 1971.

30. H. A. Wilcox, "The Ocean Food and Energy Farm Project," paper presented at the 141st Annual Meeting of American Association for the Advancement of Science, Jan. 29, 1975.

31. R. H. Williams, "The Greenhouse Effect for Ocean-Based Solar Energy Systems," unpublished working paper no. 21, Center for Environmental Studies, Princeton University.

32. Mass tree mortality may result from forest fires or plant disease epidemics.

33. Work at United Technology Corporation and Davy Powergas suggests that 60 tons of dry matter per acre per year may be feasible with water hyacinths in parts of the Southeastern United States. See R. P. Lecuyer and J. H. Marten, "An Economic Assessment of Fuelgas from Water Hyacinths," paper presented at the First IGT Symposium on Clean Fuels from Biomass Sewage, Urban Refuse, and Agricultural Wastes, Orlando, Fla., Jan. 1976.

34. Conversion of agricultural residues to methane could be an important relatively low capital cost technology for supplying energy for agricultural development in many poor countries. See A. B. Makhijani and A. D. Poole, *Energy and Agricultural Development in the Third World* (Cambridge, Mass.: Ballinger Press, 1975). Also a study by the National Academy of Sciences for U.S. AID on the application of biogasification to developing nations is nearing completion. Other conversion technologies are also being considered.

Over 95 percent of the energy demand in the United States is presently met through the combusion of fossil fuels. Although there is some disagreement regarding the exact amount of the remaining crude oil and natural gas reserves, most experts believe that the United States will become heavily dependent on synthetic fuels for its energy needs within a few decades. Coal, oil shale, and tar sands are often mentioned as attractive feedstocks for the production of synthetic liquid and gaseous fuels. Organic solid wastes, wood, and other biomass resources discussed in the Poole and Williams essay represent another source of synthetic fuels. Organic matter is a richer fuel than either oil shale or tar sands and compares favorably with coal[1] (see table 10.1). Clearly, biomass resources should not be overlooked as a feedstock for the production of synthetic fuels.

Unfortunately, organic wastes are a limited resource and attempts to increase their magnitude are restricted by the low efficiency of photosynthesis. Moreover, in the conversion of organic matter to a more useful form (for example, methane), 50 percent of the original fuel value is usually lost. The conclusion to be drawn is that photosynthetic processes result in only fairly limited quantities of net useful energy. The technology described in this article, however, overcomes these limitations by producing useful fuels (hydrogen or methanol), and by actually enhancing the energy content of the organic matter. This is accomplished by using a solar furnace to drive endothermic (that is, heat absorbing) fuel producing reactions. Thus, the fuel produced from organic wastes serves as a medium for storing solar energy. When used in this way, organic wastes become a truly major energy source.

Hydrogen has been enthusiastically proposed as the synthetic "fuel of the future."[2] As a fuel, hydrogen has many attractive properties: it is clean burning, easy to transport, useful as a chemical feedstock, very light in weight, and, contrary to popular opinion, is relatively safe. Many believe that hydrogen could replace natural gas (methane) as our primary gaseous fuel if it could be manufactured at a competitive price.

Today, hydrogen is manufactured by the steam reforming of methane $(CH_4)$:

$$CH_4 + 2H_2O \rightarrow CO_2 + 4H_2.$$

This reaction is endothermic and requires a source of high temperature heat. The heat is usually supplied by burning part of the methane feedstock and transferring the liberated heat to the steam reforming reaction.

# 10 Tower Power: Producing Fuels from Solar Energy
## Michael J. Antal, Jr.

**Table 10.1**
Energy content of raw (unprocessed) fuels

|  | Energy content (millions of Btu per ton) |
|---|---|
| Crude oil | 38 |
| Coal | 26 |
| Dry organic solid wastes | 17 |
| Wood | 15 |
| Peat | 14 |
| Oil shale | 4 |
| Tar sands | 4 |

**Figure 10.1**
The solar furnace as a tool for amplifying the useful energy available from organic wastes.

Michael J. Antal, Jr.

Clearly some other source of hydrogen will have to be found before it can supplant methane as a gaseous fuel.

Organic wastes represent a potential feedstock for the production of hydrogen. Represented by the general formula $C_x H_y O_z$ (for example, cellulose is $C_6 H_{10} O_5$), organic matter can be reformed by steam to produce hydrogen and carbon dioxide:

$$C_x H_y O_z + (2x-z)H_2O \rightarrow xCO_2 + [(y/2) + (2x-z)]H_2.$$

This reforming reaction is also endothermic and again requires a source of high temperature heat. In this report I propose that power towers be used to provide the chemical heat of reaction and produce hydrogen. The power tower is a solar furnace that uses many flat, individually guided mirrors to reflect and focus the sun's light to the top of a tower, where it is converted to heat, then electricity. The heat produced by the power tower could be stored as chemical energy in the hydrogen fuel produced.

Although the focus is on hydrogen production from organic matter throughout this report, it is worthwhile to note that the proposed system is not limited to hydrogen production alone. Methanol could be produced by the reaction

$$CO_2 + 3H_2 \rightarrow CH_3OH + H_2O$$

and used as a liquid fuel. Methane is another potential product of the system, but it is not equal to hydrogen as an energy carrier. Economic and environmental considerations will ultimately dictate which fuel is manufactured by the gasification facility.

When heated in an oxygen-free atmosphere, organic matter decomposes into a variety of gases (primarily carbon dioxide, carbon monoxide, hydrogen-2, methane and higher hydrocarbons), alcohols, liquors, oils, tars, and solid char. Although numerous technologies have been developed for manufacturing synthetic fuels from organic matter by using heat to bring about the desired chemical change or reaction (pyrolysis), the wide variety of pyrolysis products has been a major stumbling block.[3] In addition, organic wastes contain roughly 30 percent water by weight and significant quantities of heat are wasted in boiling off the water before pyrolysis of the wastes can commence.

Conversion of organic matter to hydrogen by pyrolysis in a steam atmosphere (steam reforming) avoids both of these problems. Pyrolysis products crack in the presence of steam and certan catalysts, yielding carbon dioxide and hydrogen. Moreover, the reforming reaction requires an organic waste-water mixture with a composition of roughly 55 percent water by weight and 45 percent organic matter. Thus the high moisture content of most organic wastes is advantageous for the proposed process.

The reforming reaction requires approximately 6 million Btu of heat per ton of dry organic waste. By using a solar furnace to supply this heat, the energy content of the organic waste is augmented by 36 percent of its original value. Thus with the solar furnace, one ton of dry, ash-free organic matter with a heat content of 19 million Btu is converted to hydrogen having a higher heating value of 25 million Btu,[4] which is more than three times the energy of the methane that could be produced from the same material via biogasification. This figure, together with an estimate of the magnitude and availability of the organic waste resource, can be used to project the hydrogen production potential of organic wastes. Results given in table 10.2 are based on estimates of the organic waste given by Poole and Williams, and indicate that hydrogen produced from organic waste matter could be used to meet almost half the energy demand of the United States. Thus organic matter is potentially one of the nation's richest energy resources.

## Power Towers

The idea of using a solar furnace to produce synthetic fuels by pyrolysis is not new. In his classic work *Direct Use of the Sun's Energy*, Farrington Daniels proposed that "future uses of solar furnaces might include flash pyrolysis, in which reacting systems are passed quickly through the focus of the furnace and quickly chilled.[5] Little thought, however, has heretofore been directed at fuel production by a power tower. The most obvious challenge to the union of chemical processing and power tower technologies is the difficulty in supplying the power tower's heat to the chemical reactor in an efficient and inexpensive manner.

The simplest method of delivering heat to the chemical reactor uses the heat carried by the superheated steam reactant to provide the heat of pyrolysis. Thus the steam would be heated by the power tower and piped to the chemical reactor, serving as both a reactant and a heat carrier. Unfortunately, the heat content of steam is relatively small so that more steam is required than would otherwise be necessary.

Another method for delivering heat to the reaction zone involves the location of the chemical reactor at the focus of the solar collector. Here

Michael J. Antal, Jr.

**Table 10.2**
Hydrogen production potential of wastes produced in the United States

|  | Wastes available (millions of tons) | Hydrogen-2 (quadrillions of Btu) |
|---|---|---|
| Crop residues | 340 | 6.5 |
| Manure | 45 | 0.8 |
| Plantation (nonwoody) | 750 | 14.3 |
| Urban waste | 63 | 1.4 |
| Logging waste | 100 | 1.9 |
| Pulp waste | 40 | 0.8 |
| Plantation (woody) | 300 | 5.7 |
| Total | 1,638 | 31.4 |

the reactor would serve as both a container for the reactants and a radiant energy heat exchanger. Although this configuration might lead to a more efficient use of the furnace's heat, the difficulties of operating a large chemical reactor on top of a tower should not be underestimated! It is likely that the former method will be used first to couple a solar furnace to a chemical reactor.

The proposed process would have a high overall efficiency. Theoretically, solar heat can be used to augment the organic wastes' heat of combustion by 6 million Btu per ton with perfect efficiency.[6] Realistically, heat is lost through the reactor and is carried away by the hot products, thereby lowering the process' efficiency. If we define the efficiency to be

$$\text{Efficiency} = \frac{\text{hydrogen heat of combustion out}}{\text{solar heat} + \text{heat of combustion of organic matter in}} \times 100,$$

efficiencies exceeding 70 percent appear reasonable. This is comparable to efficiencies for producing synthesis gas from coal.

There are a great many practical problems to be solved before these ideas can be implemented on a large scale. Power towers superheat steam at very high pressures, yet hydrogen production is favored by low pressures. In order to optimize power tower design for hydrogen production and minimize overall costs a far better understanding of the steam reforming reaction is required. Research presently being conducted at Princeton University and the Los Alamos Scientific Laboratory is aimed at determining optimal conditions and catalysts for hydrogen production from organic matter. It is likely that several years will pass before enough experimental

Tower Power

information can be acquired to design and construct a functional solar gasification unit.

## Economic Analysis

If a power tower pyrolysis facility were to be located in the Midwest where crop residues are plentiful, would hydrogen production from crop residues be economical?[7]

For a gasification plant processing 100 tons of residues per day, a waste shed of 70 square miles would be required. According to Poole and Williams, these residues could be delivered to the plant for $5.00 per ton. Gasification of 100 tons of crop residues is estimated to consume 120,000 kilowatt hours of heat. During an average six-hour period while the sun shines, the average yearly insolation received by the heliostat is about 0.6 kilowatts (thermal) per square meter. With a 56 percent thermal efficiency the power tower yields 0.34 kilowatts (thermal) per square meter of heliostat mirror area. Therefore, gasification of 100 tons of organic residue is estimated to require a furnace with 59,000 square meters of heliostats. The furnace and associated facilities would cover 30 acres, which is negligible compared with the 70 square miles (44,800 acres) required to generate the residues. Assuming the heliostats cost $62.00 per square meter, the heliostat array is projected to cost $3.7 million and the tower is estimated to cost $1.4 million. Including direct and indirect costs, therefore, the total cost of the power tower is estimated to be $7.4 million.

These capital costs, together with the costs of the chemical processing equipment, are shown in table 10.3. It is interesting to note that the power tower comprises roughly 50 percent of the overall costs. Operating and fixed costs are also given in the table. Interest on debt, depreciation, taxes, and after tax profit are the largest contributors to the yearly operating and fixed costs of the facility. Using these estimates, hydrogen is projected to cost $4.85 per million Btu. Although this is roughly twice as expensive as intrastate natural gas, if compares favorably with the present cost of hydrogen produced from methane by steam reforming.

Previous research[8] has focused on the production of hydrogen from municipal wastes. Economic projections for a gasification facility sized to the needs of Los Alamos, a city of some 17,000 residents, indicate that hydrogen could be produced for costs ranging from $2.90 to $3.85 per million Btu. The range was due to the large uncertainty in the estimated cost

Michael J. Antal, Jr.

**Table 10.3**
Economic analysis

|  | Costs |
| --- | --- |
| Capital costs (millions of dollars) | |
| Heliostats | 3.7 |
| Tower | 1.4 |
| Direct and indirect costs | 2.3 |
| Pyrolysis unit[a] | 2.0 |
| Compressors[b] | 0.9 |
| Shift reactors[c] | 1.3 |
| Scrubber[d] | 3.0 |
| Working capital | 1.0 |
| Total | 15.6 |
| Operating and fixed costs (millions of dollars per year) | |
| Crop residues[e] | 0.18 |
| Labor[f] | 0.10 |
| Maintenance[g] | 0.44 |
| Compressor[h] | 0.18 |
| Interest on debt[i] | 0.62 |
| After tax profit[j] | 0.62 |
| Property taxes and[k] insurance | 0.29 |
| Depreciation[l] | 0.58 |
| Federal, state, and local taxes[m] | 0.58 |
| Total | 3.59 |

Note: Yearly hydrogen production = 740 billion Btu; cost per million Btu = $4.85.
[a] Estimated by Battelle laboratories for a town of 20,000 people. [b] Compress the gas to 275 psia. [c] Remove carbon monoxide from the product gas stream. [d] Remove carbon dioxide by the hot potassium carbonate system. [e] Costs as indicated by Poole and Williams. [f] Two men present at the facility 24 hours per day. [g] Three percent per year fixed investment. [h] Electricity for compressor. [i] Eight percent per year on debt, where the debt is 50 percent of the capital investment. [j] Eight percent of the stockholder's investment, which is 50 percent of the capital investment. [k] Two percent of the capital investment. [l] Straight line basis over 25 years. [m] Fifty percent of taxable income.

of a power tower. More recent power tower cost estimates, corresponding to the midrange of these hydrogen costs, are used in this analysis.

Costs may be significantly reduced if the steam reforming reaction can be carried out at lower temperatures.[9] Less expensive solar furnaces are able to supply heat below 500° Celsius, and use of these furnaces could reduce costs below those indicated in table 10.3. Economies of scale could also be used to reduce costs. A central chemical processing facility for carbon dioxide removal and hydrogen-2 compression surrounded by several power tower gasification units with their associated biomass farms would emphasize the economies of scale associated with gas processing without detracting from the inherent advantages of the biomass farm.

Costs might also be reduced if crop residues are converted to hydrogen to meet local energy needs by a publicly owned instead of an investor owned utility (see table 10.4). The community could issue municipal bonds, say, in the amount of $15.6 million at 7 percent interest with an amortization period of 25 years. Principal plus interest payments on the bonds would cost $1.32 million, and hydrogen could be sold for $3.00 per million Btu. For the sake of comparison, intrastate natural gas recently cost $2.35 per million Btu. If the gasification facility were financed in this manner, hydrogen could nearly compete with natural gas as a fuel at today's prices.

## Conclusions

In this article I have examined the use of power tower technologies for the production of synthetic fuels. These fuels serve as a medium for storing solar energy and can be produced with high efficiency. By using a solar

Table 10.4
Operating and fixed costs for a municipal facility (millions of dollars)

|  | Costs |
| --- | --- |
| Crop residues | 0.18 |
| Labor | 0.10 |
| Maintenance | 0.44 |
| Compressor | 0.18 |
| Principal and 7 percent interest | 1.32 |
| Total | $2.22 |

Note: Yearly hydrogen production = 740 billion Btu; cost per million Btu = $3.00.

Michael J. Antal, Jr.

furnace to process the wastes and make hydrogen, the useful energy available from the organic wastes is greatly amplified. Economic projections suggest that the fuel can be produced by municipal utilities for a relatively modest price.

Organic matter serves as the raw material for producing the hydrogen fuel via steam pyrolysis. This raw material is readily available in the waste products of civilization, and a significant fraction of the nation's natural gas demand could be met by this resource alone. The biomass plantation concept advocated by Poole and Williams could be used to supply enough additional organic matter to meet the nation's entire natural gas demand.

Developing countries may find the technology described here particularly attractive. Instead of burning crop residues for heat in those countries, solar waste gasification could amplify the crop residue resource and make clean fuel available in remote locations. Thus the technologies described in this report appear suitable to both developed and developing countries where organic wastes are plentiful.

The organic matter used in this process is one of mankind's major renewable resources. Nature produces the organic matter every year as part of the carbon cycle.[10] Intelligent use of nature's carbon cycle can provide mankind with the energy it requires without desecrating the environment or leaving a heritage of radioactive wastes to generations yet unborn. The key to the success of future energy technologies lies in blending those technologies with the technologies nature has evolved over the millennia. By integrating our technologies into the natural scheme of things, ample amounts of energy will be available to sustain ourselves and generations to come.

## Notes

1. See M. Antal, "A Comparison of Coal and Biomass as Feedstocks for Synthetic Fuel Production," Proceedings of the Miami International Conference on Alternative Energy Sources, University of Florida, Miami, Fla., December 1977.

2. D. Gregory, "The Hydrogen Economy," Scientific American, 228:1 (1973), 13.

3. For example, see C. Finney and D. Garret, "The Flash Pyrolysis of Solid Wastes," Energy Sources 1:3 (1974).

4. Although it is customary to rate the energy content of fuels on the basis of their higher heating value, it may not always be feasible to make use of the latent heat of vaporization of the steam released by the combustion of the hydrogen fuel. Using the lower heating value, hydrogen produced from a ton of ash-free, organic wastes has an energy content of 20 million Btu.

5. Farrington Daniels, Direct Use of the Sun's Energy (New York: Ballantine, 1975).

6. The free energy carried by the hydrogen and available for useful work is also augmented, when compared to that of the initial waste mass, but the efficiency of conversion is less than 50 percent.

7. It is emphasized that this analysis attempts only a rough estimate of the final cost of the hydrogen fuel. More accurate data is required before reliable cost projections can be made.

8. See M. Antal, R. Feber, and M. Tinkle, "Synthetic Fuels from Solid Wastes and Solar Energy," Proceedings of the World Hydrogen Energy Conference, University of Florida, Miami, Fla., March 1976, for a more technical description of this research.

9. See M. Antal, "The Conversion of Urban Wastes," Proceedings of the International Symposium on Alternative Energy Sources, Barcelona, Spain, October 1977.

10. See M. Antal, "Hydrogen and Food Production from Nuclear Heat and Municipal Wastes," in *Hydrogen Energy Part A*, ed. T. N. Veziroglu (New York: Plenum, 1975), pp. 331–338 for a novel use of the carbon cycle.

Michael J. Antal, Jr.

# V  Development Strategies

The people of the Third World fulfill their energy needs—to the extent that they can—primarily, and in many cases almost exclusively, by the use of solar energy. This use of solar energy is dominated by photosynthetic conversion. The energy captured through agriculture in crops and crop residues provides food for people and fodder for draft animals. In some areas dung and crop residues are used for cooking. Wood and other vegetable matter constitute the principal fuels for cooking and heating.[1-3] In the past two decades the traditional energy sources have been supplemented by the use of oil, coal and electricity in industry, agriculture, transport and, to a minor extent, in homes. But solar energy remains the principal energy source for most of the people of the Third World, accounting for 60 to 90 percent of total energy use.

The solutions to energy problems of the poor are a critical part of development policy. Energy policy, like development policy, must be based on approaches that make meeting the needs of poor people today consonant with investing in the future. Because of the shortage of capital and the need to create large numbers of jobs of high productivity and quality (job satisfaction) in ways that are compatible with long term ecological realities, the poor countries cannot blindly adopt policies of imitative industrialization. Rather they must choose policies that would directly serve indigenous human needs. In particular they need to give high priorities to creating jobs in the countryside. They need to make effective use of rural resources and make their technologies compatible with them.

By adopting a radically different approach to development policy, poor countries could even set an example for the industrialized countries. Technologies and economic policies that are based on meeting human needs rather than on a free market economist's concept of demand should, in my opinion, be the foundation of energy policy for not only the Third World but for all countries.

Each human need has two aspects—the physical and the spiritual. We need houses for shelter and homes for emotional and intellectual strength and integrity. We need jobs to provide for physical needs and we need job satisfaction.

Physical needs are easily listed: food, water, shelter (including protection from extremes of weather), clothing, sanitation, and light. (The energy needed to meet these needs is implicitly included.) These physical needs are related to the needs for productive and meaningful jobs, personal mobility, physical and mental health, and education, which in turn are re-

# 11 Solar Energy and Rural Development for the Third World
## Arjun Makhijani

lated to each other. I will try to examine the role of solar energy in meeting these pressing needs.

Energy use in the Third World is much greater than is usually supposed. When food, fodder, and wood are included, annual energy per capita use is 20 to 40 million Btu rather than the 5 to 15 million Btu per capita estimated if only the so-called "commercial" energy sources (oil, coal, hydropower, etc.) are included.[1,4] This is compared with the 150 to 200 million Btu per capita that is typical of Western Europe, which, of course, is much colder.

The energy needs of the poor are not met only partly as a result of an inadequate level of energy use. The small amount of useful energy that is obtained from the energy that is used is equally important. As often as not, meeting human needs in the rural Third World involves extracting more useful work from the energy that is now used. Some typical numbers for energy use in poor countries are presented in table 11.1.

Except for food and sunshine, energy is not a need in itself. When I speak of "energy needs," it is shorthand for "the energy needed to produce and equitably distribute the wherewithal to meet human needs." In this sense, the quantity of energy needed can vary from time to time, with different forms of social organization, with different sources of energy and technologies for their use (figure 11.1).

The principal energy needs that correspond to human needs are

● agricultural fuel needs (irrigation, draft power, fertilizers, manufacture of implements, crop processing, food transport, food storage);
● energy for cooking;
● energy for providing clean domestic water supply, which, in some places, includes energy for boiling drinking water;
● house heating and warm water for bathing in cold climates;
● hot water and soap for washing clothes;
● energy for lighting (household and community);
● energy for personal transport;
● energy for processing and fabricating materials needed for house construction, pots and pans, clothes, tools, bicycles, etc.;
● energy for transport of goods; and
● energy needed to run local health services, schools, government and other community uses.

For most of the rural Third World, most of these energy needs can be fulfilled in two principal ways: (1) making the use of photosynthetic energy sources such as food, wood, crop residues, and cow dung more effi-

Arjun Makhijani

**Table 11.1**
A rough sketch of energy use in a few areas of the third world

| Principal sources: wood, food, crop residues, grazing land (except north Mexico) | India (east Gangetic plain) | China (south central) | Tanzania | Nigeria | Mexico (north) |
|---|---|---|---|---|---|
| **Domestic energy use per capita (million Btu per year)** | | | | | |
| Useful energy | 0.2 | 1.0 | 1.1 | 0.75 | 1.6 |
| Energy input | 4.0 | 20.0 | 22.0 | 15.0 | 17.0 |
| **Agricultural energy use: farm work, irrigation, chemical fertilizers (million Btu per year)** | | | | | |
| Per capita | | | | | |
| Useful energy | 0.5 | 1.4 | 0.06 | 0.16 | 13.5 |
| Energy input | 7.7 | 8.3 | 2.3 | 2.4 | 41.0 |
| Per acre | | | | | |
| Useful energy | 0.7 | 2.7 | 0.04 | 0.2 | 6.2 |
| Energy input | 10.7 | 17.3 | 1.6 | 3.0 | 19.0 |
| **Energy use per capita in transportation, crop processing and other activities (million Btu per year)** | | | | | |
| Useful energy | 0.1 | 0.1 | 0.02 | 0.03 | 0.1 |
| Energy input | 3.4 | 3.2 | 0.7 | 0.9 | 3.6 |
| **Subtotal per capita energy use (million Btu per year)** | | | | | |
| Useful energy | 0.8 | 2.5 | 1.2 | 0.9 | 15.2 |
| Energy input | 14.7 | 31.5 | 25.0 | 18.5 | 61.6 |
| **Commercial energy sources Oil, coal, hydro, etc. (million Btu per year per capita)** | | | | | |
| Useful energy | 0.5 | 1.6 | 0.2 | 0.14 | 3.0 |
| Energy input | 2.5 | 8.0 | 1.0 | 0.7 | 15.0 |
| **Total energy use (million Btu per year per capita)** | | | | | |
| Useful energy | 1.3 | 4.1 | 1.4 | 1.14 | 18.2 |
| Energy input | 17.2 | 39.5 | 26.0 | 19.2 | 75.6 |

Source: Makhijani and Poole[1] and *World Energy Supplies*.[4]
Note: These numbers are rough estimates, particularly with regard to the breakdown of animal labor into field and nonfield activities.

**Figure 11.1**
Energy to fill the most basic human requirements.

cient, and (2) increasing the photosynthetic fuel supply by increasing agricultural productivity, reforestation and planting village woodlots. In many
areas wind energy (for irrigation and crop processing) and solar water heating (to provide safe drinking water and warm water for washing clothes)
can provide supplementary energy which the rural people can afford. Adopting this human approach to providing energy needs is also important for
initiating soil and water conservation programs which are important to
maintaining and increasing the productivity of the rural environment.

### Agricultural Energy Needs

Agriculture is essentially the organized trapping of solar energy in crops
and crop residues. In general, the energy content of the crops plus crop
residues exceeds the energy inputs to farming, including energy for irrigation, manufacture of chemical fertilizers, and the energy to make and
power farm machines (excluding crop transport, storage and cooking).
This applies to all grain farming systems that I have studied, ranging from
farms on which only manual labor is used to farms assisted by animals to

Arjun Makhijani

completely mechanized farms. Where inputs, including energy, to agriculture are increased properly the energy output per unit of land increases, that is, the plants capture solar energy more efficiently.

As an example, when the energy inputs and outputs of two prototypical villages (one in Bihar, India, and the other in northern Mexico) were compared, I found that the net energy output of food and crop residues increases much faster than the energy inputs of irrigation, fertilizers, and double cropping.[1] Increasing annual energy input from 11 million Btu per acre to 20 million Btu per acre increases the energy output from 16 million Btu per acre to 48 million Btu per acre, raising the efficiency of solar energy capture from about 0.07 percent to about 0.2 percent. The most important feature of the increase in energy inputs is that annual useful energy input is increased from 0.8 million Btu per acre to 6 million Btu per acre.

The principal energy needs in agriculture are related to increases in (1) the number of jobs; (2) the output per worker; and (3) the output per unit of land (that is, increasing the solar energy capture in agriculture).

To accomplish large increases in employment and productivity, fertilizers and irrigation are usually essential. Either of these or both in combination can help achieve simultaneous increases in all three factors. This is reinforced by the fact that the warm climates prevalent in these areas permit harvesting several crops a year, though the (erratic) rainfall is largely limited to a single monsoon season.

Planting legumes or seeding the soil with nitrogen-fixing bacteria are ways to use solar energy to increase the stock of fixed nitrogen in the agricultural economy. The mechanical energy that is generally needed for irrigation from underground water sources during the dry season can be provided either from biogasification or from wind energy in many instances, and perhaps in most cases more economically than with diesel or electricity.

Total fuel requirements for irrigation and fertilizers can vary from 2 to 20 million Btu per acre per year, depending on the number of crops, method of irrigation, source of water, source of energy, etc.

The other major requirement for energy in the fields is draft energy. This is usually provided by human and animal labor and in some places by tractors. A characteristic feature of energy requirements for field operations is that they peak sharply at certain times of the year (plowing, planting, harvesting, and, in some instance, crop processing). Thus, many farmers experience acute shortages of human and animal labor during brief periods while many are unemployed for a good part of the year. In certain areas, such as the Deccan plateau of India, there is a severe lack of power to till the fields because the hard soil may require 2 to 3 horsepower to turn the soil effectively. This is more than the one-half to one horsepower available even to a farmer with a medium size plot (5 to 20 acres of dryland).

In some areas shortages of draft power and labor may be a factor in preventing the adoption of productive but labor intensive operations such as rice transplantation. In such areas increasing draft animal power or selective mechanization can help increase output. However, this can cause a large loss of jobs for the landless unless other measures are taken prior to and during the mechanization period.

Mechanization cannot succeed in alleviating the problems of the poor where a small minority controls most of the land and there are large numbers of landless people. The organization of labor unions and sharing of animals can considerably ease the labor shortages that small farmers commonly experience. Such organizations would also guarantee an increase in employment for the landless and should precede efforts at providing machines to solve peak labor problems. In some places such efforts can significantly postpone the need for selective mechanization while increasing both output and employment.[5] In any event, in those cases where mechanization would endanger jobs, particularly those of the landless and insecure tenants, mechanization must be postponed until measures are taken to ensure it will not. As Gandhi said in 1973, machines must not be pitted against human beings.

Windmills can be practical in some areas for water pumping, crop processing, and rural industries for which the energy requirements can be spread out over a number of days. Since there is no need to supply energy instantly, the major costs of conversion of mechanical energy to electricity (or some other form), storage and reconversion to mechanical (or other forms) are avoided. The basic criteria for windmill design and their relation to the values discussed are as follows:

• The total cost should be low; it should be less than $200 and preferably less than $100 so that small farmers and others who may need a source of power for cottage industries would be able to afford it. (These costs include the cost of labor.)
• It should have sufficient power to enable irrigation of a small farmer's plot (0.5 to 1 acre). The size of the windmill, therefore, depends on the local wind conditions and on the source and quantity of water needed.
• It should be made of local materials as far as possible and with local skills (that is, those available in a village). Among the many reasons for establishing this criterion are (a) low capital input and high labor intensity;

Arjun Makhijani

(b) accessibility of the technology to the poor; (c) ease and quickness of maintenance and repair when required; and (d) high job quality in terms of satisfaction in one's work and control of it.

• It should be stable under adverse weather conditions.

• It should have high starting torque so that auxiliary starting devices are not needed.

Work on a windmill meeting these design criteria is now under way at the Indian Institute of Science in Bangalore. A vertical axis windmill with a Savonius-type rotor was chosen to meet the criteria of low cost and high starting torque. To allow fabrication with local materials the rotor will consist of cloth sails stiffened by ropes and attached to circular boards at the top and bottom. The choice of cloth and ropes as the materials for the rotor will also enable the sails to be shaped to achieve high efficiency and stability. Several models were tested in the Institute's wind tunnel and a rotor with tapered sails was selected. Tapered sails entail some loss of efficiency but enhance the stability of the windmill. The efficiency of this windmill has been estimated to be roughly 20 percent. A prototype with 10-foot high rotor will be tested this year to determine the power output characteristics more definitely. It is anticipated that the only part needed for the windmill that cannot be made in a village will be the bearing. A rough sketch of this windmill is shown in figure 11.2.

The production of methane from nonwoody organic matter by the action of anaerobic bacteria (biogasification) could be among the most important energy technologies for the Third World. The technical details of biogasification have been discussed rather widely.[1,6] Therefore, apart from a brief discussion, I will confine myself to applications (particularly in agriculture) and a few implications for energy policy.

Anaerobic bacteria acting on nonwoody organic matter produce a gas which is principally a mixture of methane (60 to 65 percent by volume) and carbon dioxide. Of the trace gases, hydrogen sulfide is the one which concerns us the most, but it is easily removed. The residue (liquid with solids in suspension) is a rich source of fertilizer and humus. Maintenance of proper temperature, pH and carbon to nitrogen ratios is important to efficient and smooth operation of biogas plants.

Community-size biogas plants (total capacity greater than about 1,000 cubic feet of biogas per day) can be an important and economical source of fuel and fertilizers for agriculture and particularly for irrigation. Stationary gas, gasoline and diesel engines as well as mobile gas and gasoline

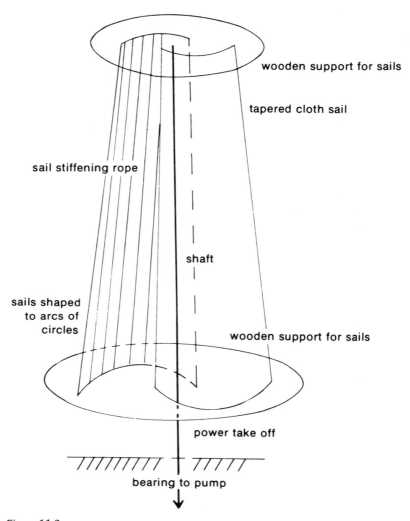

wooden support for sails

tapered cloth sail

sail stiffening rope

shaft

sails shaped
to arcs of
circles

wooden support for sails

power take off

bearing to pump

Figure 11.2
Sketch of modified Savonius rotor windmill.
Source: R. Narasimha, Indian Institute of Science, Bangalore, India.

Arjun Makhijani

engines can be easily and cheaply adapted to the use of biogas or methane. Methane is obtained when the carbon dioxide is removed from biogas, for example, by passing it through a lime water solution. In both cases the trace quantities of hydrogen sulfide must be removed (usually by passing the gas through a bed of iron filings) to prevent corrosion.

The production of biogas for agricultural uses has to be closely matched with the anticipated seasonal fluctuations in demand. It is prohibitively expensive to store large quantities of biogas or methane.[7] The rate of production of gas can be controlled by loading as many of the digestors as will be needed for the succeeding period of one to two months. The digestors in this scheme can be of the batch type so that a variety of organic materials can be loaded simultaneously, provided the combination has the right carbon to nitrogen ratio. This ratio will be too high if crop residues are mixed with dung. It can be decreased by addition of human excrement, including urine, and chemical nitrogen.

The digestors that are not in use would store the residuum (fertilizer), and the spare gas holders could be used to store gas. Crop residues would be stored for loading the digestors just prior to periods of peak demand. Human excrement would be used at an approximately constant rate throughout the year. In times of low gas use, this would be supplemented by some animal dung and some crop residues (to achieve the right carbon to nitrogen ratio). In times of peak use all the available dung and a good portion of the crop residues could be fed to the digestor.

The cost of compressed methane in a village scale community biogas plant installation works out to about $4.50 per million Btu (table 11.2). (This cost estimate may well be high. Other estimates of the capital costs for the biogas plant and storage cylinders range down to about a quarter of what I have assumed here.) This is somewhat less than the cost of diesel excluding taxes, but including about $1 per million Btu transport charge. In terms of useful energy it is comparable to electricity at about 6 cents per kilowatt hour and therefore would be generally cheaper than unsubsidized electricity. The foreign exchange requirements for most Third World countries would be much smaller in the case of biogas than either diesel or electricity. The same can also be said of the total capital requirements. Further, jobs would be created in the countryside—where they are needed. The number of jobs would be 10 to 100 times greater than in a centralized electricity and fertilizer production scheme.[8]

As with all development efforts, organizing for building and maintaining community biogas plants is essential to ensure proper maintenance

**Table 11.2**
Cost of a community biogas scheme for irrigation

|  | Dollars |
|---|---|
| Capital costs | |
| Digestors and gas holders (400 cubic meters per day capacity)[a] | 10,000 |
| Compressor and engine | 2,000 |
| Land cost | 500 |
| Extension services, etc. | 1,500 |
| 40 cylinders (0.03 cubic meter capacity for irrigation fuel distribution)[b] | 5,000 |
| Total | 19,000 |
| Annual costs | |
| Interest on capital | 2,280 |
| Depreciation (5 percent) | 950 |
| Cost of dung and crop residues (at $2.00 per ton)[c] | 800 |
| Labor and maintenance | 1,000 |
| Subtotal | 5,030 |
| Sale of gas plant residues on the basis of nitrogen content ($400 per ton nitrogen) | 1,000 |
| Net annual cost for 1 billion Btu of gas | 4,030 |
| Cost per million Btu of compressed gas about[d] | 4.50 |

[a]400 cubic meters per day capacity enables production of about 1 billion Btu of methane per year (150 cubic meters of biogas per day) on a schedule compatible with irrigation needs for a 3 crop per year cycle. The energy could irrigate from 25 to 250 acres depending on local conditions.
[b]0.03 cubic meters methane at 200 kilogram per square centimeter is enough to power a 5 horsepower pump for 4 hours.
[c]70 percent assumed to be dung and human excrement, 30 percent crop residues. Not all the available dung and human excrement can be used because gas output fluctuates seasonally. The dung and excrement not used are assumed to be composted.
[d]10 percent of the gas is used for compression.

Arjun Makhijani

and equitable distribution of the benefits. Widespread implementation may also require that some of the capital be raised in the villages themselves, which are now so poor. To meet these challenges of organization and capital requirements, proper timing of the efforts is necessary. Initiating composting schemes could be one effective way both to organize and raise resources for building community biogas plants.

A comprehensive composting program could increase annual food production in poor countries by over 100 million tons—about 4 times the current deficit in terms of need. Additional nitrogen fertilizer on the order of 10 to 15 million tons per year could be had if only half the available nitrogen in human and animal excrement is used. (China has been excluded from these calculations.) This has the potential of increasing to 20 to 25 million tons as diets improve (that is, protein intake increases), as recovery of wastes is increased, and as biogas plants are used to conserve nitrogen. In addition, roughly 5 million tons each of $P_2O_5$ (phosphorus pentoxide) and $K_2O$ (potassium oxide) would be available in the compost. These figures can boost existing use of chemical fertilizers significantly. In 1971, chemical fertilizer use in poor countries was 7 million tons of nitrogen, 2.7 million tons of phosphorus pentoxide, and 1.5 million tons of potassium oxide. In the Gangetic plain, Iraq, Ethiopia, and some other regions, composting programs must be tied to the reduction of dung for cooking.

## Cooking/Heating Needs

Typically, the annual use of wood or any of the other photosynthetic fuels for cooking is about 5 to 7 million Btu per capita (about 2 pounds of dry wood per capita per day). This is considerably larger than the 3 million Btu per capita in U.S. gas stoves and ovens, excluding the pilot light fuel use. The open fires and stoves currently used for cooking are cheap, but generally inefficient.

In regions of high altitude, desert regions and areas outside the 20°N–20°S latitude belt, winters are cold and people need to keep warm. Where fuel wood is available people use large quantities of it to keep warm. Where it is not (for example, the Indo-Gangetic plain) and where the poor do not have enough clothes, blankets or adequate beds to keep warm, they suffer the cold and some die of it.

Success in meeting domestic energy needs and in soil and water conservation programs therefore involves the following considerations:

Solar Energy and the Third World

• Reducing the use of wood, crop residues, and dung as fuels as soon as possible while meeting the requirements of cooking and keeping warm. This must be done on a broad front and must include the rural and urban poor. The technologies involved must, therefore, be of low cost and high labor content (so people can do it themselves).

• Large scale reforestation programs.

• Organizing fuel wood supply for entire villages so that the cutting can be done within minimum damage to the soil and can be accompanied by replanting efforts.

• Organizing and planting village wood lots so that within a 10 or 15 year period the necessity for cutting trees in areas outside the village is entirely eliminated.

Reforestation programs can be a source of employment for large numbers of people,[9] including those who will lose income currently obtained from the sale of wood. Many poor people in all poor countries depend on gathering wood in the forests for sale in villages, towns, or cities.

Commercial exploitation of forest resources must be accompanied by replanting and controlled effectively to ensure continued forest productivity. Blatant disregard for forest productivity and ecological sanity is a common feature of many large commercial operations, perhaps best typified by the clear cutting of Indonesian forests by Japanese industries.

Technically speaking, the use of wood, crop residues, and dung can be reduced in three ways: (1) fuel substitution; (2) converting wood, crop residues or dung to some other form so that they can be used much more efficiently; and (3) making the direct use of wood, crop residues, and dung much more efficient.

For several reasons, principally economic, the third approach is the only widely applicable one. Thus, for example, the Indian government's program of building family size biogas plants for providing cooking fuel is simply beyond the reach of most people since each plant costs $200 and little productive use is made of it. At most, the program will reach 1 percent of India's rural population after 10 years. Kerosene, of course, is prohibitively expensive; coal, peat, and similar fuels may help in some areas, but they are not accessible to most Third World people and, even when they are, they are often too expensive for the poor.

To reduce the use of wood, crop residues, and dung significantly in the next 10 to 15 years, it will be necessary to vastly increase the efficiency with which these fuels are used for cooking; to provide for warmth

Arjun Makhijani

largely through improved clothing, blankets, beds, and houses; and to supplement the use of these fuels with efficient heating systems in those areas which experience severe cold.

In order that efficient cooking stoves and methods be rapidly adopted, they should be cheap and made with local materials. In order to translate these criteria to the realm of practice, I will discuss them with specific reference to conditions in rural India.

Typically, a family of five uses 1.5 to 2 tons of wood (or an equivalent quantity of some mix of dung and crop residues) per year. For those who purchase wood, the price ranges from $20 to $50 per family ($4 to $10 per capita). For those who are too poor to purchase wood—and those people are probably the majority—gathering fuelwood may involve anywhere from 50 to 200 or more days of work per family, depending on the accessibility of wood supplies.

For a stove to be "cheap," it must cost less than the annual cost of the wood that would be saved (capital to output ratio less than 1). For those who have no cash, it should save them more labor in a year than they would have to put into purchasing the stove. Therefore, if a stove cuts fuel consumption in half, its cost should be less than $25, and preferably less than $10. Alternatively, building a stove should require less than 20 to 30 days of work.

It appears that the principal defects of the cooking stove used in rural India are the lack of adequate air supply and draft control. This results in a smoky low temperature flame which is both inefficient and unhealthy. The cost of a two "burner" stove is $1 to $2. The principal defects of the stove can probably be rectified by the addition of a chimney and an additional air inlet with a damper. The chimney can be made of clay pipe.

Considerable testing on the efficiency and cost of various burner, damper and chimney configurations is needed; but even with the designs in hand, the total cost of the stove should be under $10. These conjectures will soon be tested in the field in Majarashtra, India.[10] Some testing has already been undertaken by the Hyderabad Engineering Research Laboratories in Hyderabad, India, which developed a cooking stove with a water heater. A reduction of 25 percent in fuel use is claimed for this stove (without different cooking methods), and it provides hot water as well.

## Keeping Warm

The most immediate need for keeping warm in many areas of poor countries is adequate clothing. The poorest people though usually have only

Solar Energy and the Third World

**Figure 11.3**
Research must draw on the realities of the countryside and the experience of rural people.

one set of clothing which they wear all the time, and the children often have little or none. The life of the clothing is also greatly reduced because they are washed in cold water by beating against rocks and often without soap. There are, however, four devices that complement each other and can help solve the problem of adequate clothing:

- the hand operated spinning wheel (Gandhi's *Charkha*),
- the handloom,
- a simple solar water heater, and
- a hand operated washing machine.

Warm water for washing clothes can be obtained from a simple solar water heater. This could consist of a wooden tub (as wood is a poor heat conductor), which has been painted black on the inside and covered with a sheet of glass. The hot water would be used in a hand-operated washing machine.

It would be more expensive to operate the scheme using wood or any other fuel: at 30 percent boiler efficiency and $15 per ton of wood, the

Arjun Makhijani

annual fuel cost alone would equal the total solar water heater cost. Using wood would also require at least 30 times the land area needed by the solar heater. However, it may be desirable to integrate hot water production with the cooking stove, as this may be cheaper and does not require additional fuel.

The cost of these simple solar water heaters is quite high. The more conventional solar water heaters may be more economical. A solar water heater made with discarded fluorescent tubes may also be suitable.[11] Even so, the annual cost of warm water and a washing machine would be around 50 to 60 cents per capita. For this amount, the life of clothes would be greatly extended and time would be saved. The waste water could be used to irrigate a community vegetable garden.

In very cold areas heating is needed in addition to clothing and blankets. The people of Shensi in central China fulfill part of this need by building clay or stone caves in the hills which are partly warmed by the earth's heat. The sunlight comes in through the translucent paper which covers the arched transom over the doorway. This is not enough for the very cold winters of northern Shensi. The people keep warm by heating the surface of a bed known as *Kang*, which has a small wood fire underneath and a chimney that leads out of the cave. By warming a surface rather than the entire volume of their homes, I would guess the people of Shensi probably save enormous quantities of scarce wood fuel.

## Village Public Utility

The intense involvement of villagers in community projects, which is essential to their long term success, could be obtained by organizing a village public utility that would be responsible for the domestic water supply, sanitation, and fuel supply. Most of the funds for this project would be loaned by the development agency (government or nongovernment), though small amounts of capital can often be raised in the villages themselves.

Plans for water supply, fuel supply, the supply and installation of cooking stoves, and sanitation and composting would be prepared by a managing committee (composed of villagers) with the help of the development agency that is the source of the loan for the project. Such plans would be modified or approved by the village as a whole.

Payment by the people for the goods and services supplied by the utility would be in the form of labor. In order to pay back the loan, the utility would acquire a small parcel of land on which the labor would be used.

The compost produced from the sanitation facilities would be applied to the utility land or sold directly or some combination of the two.

A village of 1,000 people (200 families) would need 200 tons of fuelwood each year to meet cooking fuel needs. This assumes that current cooking fuel use would be cut in half by efficient stoves and cooking methods. A village woodlot would require 25 acres of land to supply this quantity of wood on a renewable basis of 8 tons per acre per year. A substantial portion of this land requirement could be fulfilled by lining the village streets, approach roads, windbreak areas as well as other areas with trees. The rest of the land could be obtained from the village commons, or it could be some other piece of land that was not currently being cultivated. Management of the woodlot by the villagers themselves would considerably reduce illegal felling of trees.

Establishing a woodlot to meet the entire fuelwood needs at the rate of about 400 pounds per year per capita will be a difficult matter in some areas such as the Gangetic plain where little land is available for establishing woodlots. Every effort must be made in such cases to use as efficient stoves and cooking methods as possible and to meet the fuel needs from the trees established on village streets, etc. Such efforts would pay for themselves in terms of the increased food production that would result from using surplus manure as a fertilizer instead of fuel.

While the woodlot is being established, the village fuelwood supply could be organized by the community. A supply of wood for several months for the entire village could be harvested during the off-peak periods. Organizing the fuelwood supply for every village in an area in this manner would allow the integration of regional afforestation and reforestation schemes with the fuel needs of the people. Since installation of efficient stoves would be one of the first actions of the villagers and the utility, the amount of labor expended in gathering fuelwood would be substantially reduced. Organizing a fuelwood supply in this way until the village woodlots come to maturity would not, so far as I can see, involve additional investments by the utility.

Coppicing—the practice of cutting trees in short rotation and allowing the sprouts to reappear from the stumps of the trees—can significantly reduce the time of maturation of a village woodlot.

Some tentative cost figures for an Indian village of 1,000 people (200 families) indicate a total initial investment of $12,000. The loan would be repaid in 5 to 7 years with interest (at 12 percent). Each family would obtain water for household use, fuelwood, a stove and (community) sanita-

Arjun Makhijani

tion facilities in return for about 35 days work (per family). Note that this is considerably less than the amount of labor which the poor spend today on gathering fuelwood alone. (It is instructive to compare this with the number of days per family that a typical U.S. family has to work to provide itself with water, sanitation, cooking fuel, and cooking stoves. If the after-tax family income is taken as $800 per family per month, an annual $500 cost—including interest charges and fuel costs—yields a figure of around 15 days work. On a per capita basis, the labor cost is about 5 days—comparable to the 7 days in the above example of an Indian village.) This work would be directed toward (1) producing one to two crops per year on 7 acres (on the order of 1 percent of the cultivated land in a typical Indian village) in order to repay the loan; (2) maintaining and managing the facilities of the utility; and (3) establishing and maintaining the village woodlot.

The schedule of work would be determined by the managing committee subject to majority approval of the village. Since traditional agricultural communities experience severe peaks and valleys in labor requirements, the work schedule to pay back the utility would probably be geared to those months which are normally idle. That is why provision should be made for irrigating the public utility agricultural land. The utility land could, of course, be used to grow more than one or two crops per year.

The allocation of the work and the distribution of benefits must be satisfactorily worked out by the people of the village as part of the organizational work involved in creating the public utility. The land could, for example, be used to provide an extra source of work and income to the landless or to provide funds which could be used for community activities, such as a school, village roads, public lighting, library, etc.

## Research

The conventional approach to energy and much other research in the industrialized countries has been to find technologies that can be "commercialized." The word "commercialized" is generally synonymous with mass production of goods in capital intensive factories. This applies to all manner of goods (and these days "services") from solar cells to automobiles to nuclear power plants. A principal objective of this approach is to take advantage of what economists call "economies of scale." This approach conditions the type of research that is done. Thus, electric power is regarded

Solar Energy and the Third World

by economists as a "natural monopoly." In fact there is nothing "natural" about it.

The fact is that the technologies we have developed to supply electricity are biased in favor of those which have large economies of scale. Hence we have a monopoly of electric power that is not "natural" but a product of biases in research.

This approach often not only ignores environmental damage (whose "costs" often cannot be measured, much less compensated for), but it creates dull jobs over which people have little control and in which they can find little satisfaction. Some of the technologies that this approach produces are irreversible in their effects. If we find as we learn more—and who would argue that we know all and have perfect foresight to boot— that they are incompatible with human needs and ecological realities, we cannot undo their effects. Nuclear power is a prime example of such a technology, and genetic engineering may soon have the dubious distinction of superceding it in this regard.

The Third World must adopt a different approach to research. The general criteria for research should be similar to the ones cited for the windmill— low cost, emphasis on local materials, high labor intensity and productivity, high job satisfaction, the capacity to meet the needs of the people within their means, and compatibility with long term environmental enhancement. Attention needs to be focused on use of technologies and their effects on income distribution. We must concentrate on what A. K. N. Reddy calls "inequality-reduction technologies."[12]

The development of these technologies should be accompanied by the finest scientific and social thinking. Rooted in the realities of the countryside and drawing on the store of knowledge and experience of rural people, such research can produce technologies that, coupled with educational and organization efforts, can deserve the adjectives "high" or "advanced." The commercialization of such technologies will not be in the building of factories but in the extension work in villages and towns.

There is some ferment in this direction in many countries. In India, the Indian Institute of Science has begun just such a research program. Much more effort is needed along these lines. Here I will list a few of the solar energy related technologies and systems which seem important in energy research.

● Harvesting tropical and subtropical bogs.[13] Plants in bogs are among the more efficient photosynthetic converters of solar energy (4 percent, that is, much more efficient than agricultural systems). Poor countries have

Arjun Makhijani

plenty of these bogs: enough to produce up to 300 quadrillion Btu of bio-mass energy (excluding the potential of floodplains) annually on a renew-able basis if the nutrients are returned to the bogs. The plant matter could be converted to methane and carbon dioxide by anaerobic digestion. The scale of operation can be adjusted to suit local needs. Research on this is particularly urgent since all over the world tropical and subtropical bogs are being filled in without thought to their potential for energy production.

• Extensive testing of cooking stoves made with local materials from all over the Third World is a critical research need.

• Synergistic systems in which food or some other needed item is the principal product, and energy, fertilizers, raw materials are byproducts. The use of the coconut palm tree and its products by the people of Kerala is a good example. The following hypothetical example may be useful in some areas: a water tank (created with an earth dam) is used to grow fish and water hyacinths; the water hyacinths are used for fuel and fertilizer via biogasification; the fish wastes are fed to chickens; the chicken manure to pigs; the pig manure to the biogas plant; and part of the biogas plant fer-tilizer output is returned to the tank.

• Small biogas plants costing less than $50, which are made with local materials, for use by village families. Human excrement and food wastes put into the biogas plants could provide enough fuel for most household uses, particularly if coupled with efficient cooking pots, stoves, and methods.

• Pilot community biogas plants with particular emphasis on involvement of people and equitable distribution of benefits. The gas should be used primarily to increase food production and jobs.

• Solar cookers. Much effort has been devoted to producing solar cookers. Unfortunately the materials in current designs are inaccessible to most Third World rural people. In principle, solar cookers could contribute sig-nificantly to reducing wood supply problems since a year's cooking energy for one person can be supplied from an area of one or two square yards.

• Using sewage gas for buses and trucks. Sewage treatment plants produce methane which can be used as a substitute for expensive oil in trucks and buses. Using purified sewage gas (methane) in this way would not only save foreign exchange, but enhance the feasibility of sewage treatment pro-jects in towns and cities all over the world. A study and a pilot testing pro-gram are needed. Compressed natural gas (mostly methane) is currently used commercially by operators of truck and automobile fleets in France, Canada, and the United States.

- Small scale hydropower pilot projects and a survey examining their potential for providing either mechanical energy directly or electricity are needed. The techniques that are presently being used in China could serve as a useful example.

This partial list of areas where research in solar energy technologies is badly needed for the Third World does not include research in centralized power production from solar energy. While poor countries should keep track of developments in this area, the capital intensiveness of the current approaches warrants a rethinking of the suitability of central station solar power for poor countries. Current energy related problems in the Third World do not arise from a lack of available technologies for using solar energy to benefit the poor; problems arise from the tendency of policymakers in poor countries to favor imitative industrialization. This results in the widespread neglect of the rural poor. Once development policymakers view their jobs as protagonists of the poor, they are not likely to confront a shortage of ways to use solar energy.

### Notes

1. Arjun Makhijani and Alan D. Poole, *Energy and Agriculture in the Third World* (Cambridge, Mass.: Ballinger, 1975), chapters 2 and 4.

2. Eric Eckholm, *Losing Ground—Environmental Stress and World Food Prospects* (Washington, D.C.: World Watch Institute, 1976).

3. Keith Openshaw, "The Gambia: A Wood Consumption Survey and Timber Trend Study, 1973–2000." Unpublished report to the ODA/LRD Gambia Land Resources Development Project, Midlothian, Great Britain, 1973.

4. United Nations, *World Energy Supplies, 1960–1970,* Statistical Papers, Series J, No. 15 (New York: United Nations, 1971).

5. Jan Myrdal, *A Report from a Chinese Village* (New York: Random House, 1965).

6. M. A. Sathianathan, *Bio-gas Achievements and Challenges,* (New Delhi: Association of Voluntary Agencies for Rural Development, 1975).

7. In this respect the calculations on the cost of biogas presented in Makhijani and Poole [1, Table 4–9] are wrong. The costs of storing six months of gas production far exceed those cited. It is much cheaper to build several digestors and control the loading rate.

8. United Nations, Food and Agriculture Organization, *Production Yearbook,* 1972, vol. 26, Rome, 1973.

9. Makhijani, "Food as Capital for Third World Development," Institute for Public Policy Alternatives, State University of New York, Albany, New York, 1975.

10. The stoves will be tested by me under the auspices of the Foundation for Research in Community Health (Dhokawade, P. O. Awas, Alibag Taluk, Kolaba District, Maharashtra, India).

11. Marshal Merriam, unpublished solar energy notes, Department of Materials Science, University of California, Berkeley, 1975.

12. A. K. N. Reddy, unpublished paper, "The Technological Roots of India's Poverty," Indian Institute of Science, Bangalore, India.

13. Alan D. Poole and Robert H. Williams, "Flower Power: Prospects for Photosynthetic Energy," *Bulletin of Atomic Scientists,* May 1976.

At present, the United States is approaching its energy and economic problems by striving to continue past trends. We believe that this is not an effective policy. Fuel conservation should be a major part of any long-term solution for these problems. It is the best energy policy for simultaneously stretching out our limited fossil fuel resources, holding down total energy costs, minimizing dependence on foreign energy sources, and protecting the environment.

It is entirely feasible for the United States to pursue something approaching zero energy growth for several decades without economic hardship, if the nation should deliberately choose this course. Further, there is good evidence to suggest that an aggressive pursuit of energy conservation may be one of the more promising roads to economic well-being over the long run, through reduced consumer expenditures for fuel and electricity, and increased job opportunities.

The estimates presented here for the potential of fuel conservation are greater than most other estimates, largely because we take into account technological innovation in fuel conservation. Most other estimates assume very little, if any, technological change. We must dispel the popular and very erroneous notion that once the "fat" in our national energy budget has been eliminated in the next few years, further efforts to hold back energy growth would "cut to the bone" and jeopardize the economy.

In support of this thesis we will bring together results of research carried out for the Energy Policy Project of the Ford Foundation, and the more recent results of an energy conservation study by the American Physical Society.[1]

We begin by illustrating the notion that fuel conservation can radically alter the energy supply situation. Consider an enthusiastic house buyer who wants solar heat. The buyer commissions a design for a standard modern house with rooftop solar collectors, a basement heat storage system, plus an inexpensive oil- or gas-fired backup system for the longer spells of cloudy winter weather. A typical optimization of the solar and backup systems might call for two-thirds of the heating from the solar unit and one-third from the backup system. Alternatively, the prospective house buyer could choose a fuel conservation design: with thermally massive walls, substantial insulation, well designed and built doors, windows, and flues, large south facing windows with summer shading,[2] sophisticated thermal controls, etc.

This house would need less than one-third the fuel for heating as a standard house. It would use no more fuel than the backup heating system in

---

## 12 Energy Efficiency: Our Most Underrated Energy Resource
## Marc H. Ross and Robert H. Williams

the house with solar collectors. The fuel conservation house would probably be cheaper—if builders were accustomed to building such houses. This
illustration is not, of course, an argument against solar heating. It, and the
analogous examples presented throughout this essay, illustrate the fact
that effective use of energy can reduce the energy needed for a given task
so significantly that provision of that energy is no longer problematic.

Energy consumption associated with any activity or process is obtained
by multiplying two factors: the demand for a product and the specific
energy (the energy required to provide each unit of the product). For
example, the fuel used to drive automobiles is the product of the number
of miles driven and the specific energy. The specific energy in this case is
the average number of gallons consumed per mile, the reciprocal of the
familiar miles per gallon. Energy can be conserved either by curbing demand or by reducing the specific energy. The demand for products depends
on considerations like personal goals, income, and the like. While changes
in demand may be desirable in some areas, such changes depend on highly
uncertain consumer attitudes. In this essay we will not speculate on how
consumers might modify their behavior in the pursuit of energy conservation goals. Instead, we focus attention primarily on conservation opportunities associated with reducing the specific energy involved in providing
goods and services, that is, on efficiency improvements.

### Efficiency Measures

We need to introduce a concept which illustrates how inefficient most energy consuming processes are, and which can be used to point up the possibilities for efficiency improvements. The energy conservation literature is
peppered with discussions of the efficiency of energy use. Unfortunately,
the efficiency concepts commonly used are entirely inadequate indicators
of the potential for fuel savings. A couple of examples will illustrate this
point.

● Household furnaces are typically described as being about 60 percent
efficient—in the sense that 60 percent of the heat of combustion of the
fuel can be delivered as useful heat to the house. This measure suggests
that the most useful heat you can get from a fuel is 100 percent of the
heat of combustion. But this is incorrect because devices exist for actually
delivering more! A heat pump, by extracting energy from the out-of-doors
and "pumping" it up to useful temperatures, can deliver as useful heat
more than 100 percent of the fuel energy it consumes.

• Air conditioners are rated by a coefficient of performance (COP), which is the heat extracted from a cooled space divided by the electrical energy consumed. A typical air conditioner might have a COP = 2 (that is, 200 percent efficient). Unfortunately this measure provides no hint as to how this performance compares to the maximum possible, which is a COP much greater than 2.

In these and other cases the efficiency is defined as the percentage of the energy input to the system which is delivered for the desired purpose. A much more meaningful efficiency measure is one which measures performance by showing for a given task how the actual fuel consumption compares to the theoretical minimum amount of fuel needed to perform the task. Because the minimum amount of fuel needed to perform a given task is set by the second law of thermodynamics this new efficiency measure has been called the "second-law efficiency."[3] The more familiar efficiency measure, described above, is called the "first-law efficiency."

To illustrate how the second-law efficiency concept works, suppose a thermodynamic calculation shows that a minimum of 1 gallon of oil is needed for a given task, but that the technology at hand required 10 gallons. The second-law efficiency is thus one-tenth or 10 percent. Only for an "ideal" system can the second-law efficiency be 1 or 100 percent. As the examples in table 12.1 indicate, the second-law efficiencies for most tasks are less than 10 percent. This clearly shows that energy is being used very inefficiently today. In particular, the household gas furnace described above has a second-law efficiency of only 5 percent. Thus, while the first-law efficiency in this example gives us the misleading impression that only a modest efficiency improvement may be possible, the second-law efficiency correctly indicates a twentyfold maximum potential gain.

Just how far can we expect to go toward achieving the theoretical maximum of 100 percent efficiency?

In practice 100 percent efficiency is never achieved. This maximum is limited by both available technology and economics. At some point the fuel savings associated with a further efficiency gain are not worth the additional cost. Our judgment, which is based on the study of a variety of devices and processes, is that over the long term a goal of 20 to 50 percent is reasonable for ultimate practical systems. The values at the high end of this range would be more characteristic of highly engineered devices designed for specialized tasks (mainly in industry), and the values at the low end of the range would be representative of what could be achieved with more flexible, less sophisticated devices suitable for wide applications in

**Table 12.1**
Second law of efficiency for typical energy consuming activities

| Sector | Second-law efficiency (percent) |
|---|---|
| **Residential/commercial** | |
| Space heating | |
| Furnace | 5.0 |
| Elective resistive | 2.5 |
| Air conditioning | 4.5 |
| Water heating | |
| Gas | 3.0 |
| Electric | 1.5 |
| Refrigeration | 4.0 |
| **Transportation** | |
| Automobile | 9.0 |
| **Industry** | |
| Electric power generation | 33.0 |
| Process steam production | 34.0 |
| Steel production | 23.0 |
| Aluminum production | 13.0 |

our homes, in buildings, and in transportation. Thus, there is considerable room for efficiency gains through innovation, starting from today's technology.

While the second-law efficiency measure suggests potentially enormous opportunities for savings, it does not tell the whole story because the efficiency given is for a specific task, which can often be modified without adversely affecting the quality of the product provided.

For example, table 12.1 indicates that the second-law efficiency of aluminum production is 13 percent. But this is the efficiency for producing aluminum from virgin ores, where the theoretical minimum energy requirement is 25 million Btu per ton of aluminum compared to the 190 million Btu actually used today. If the task is redefined to allow for recycling, the potential for fuel savings is even greater, since aluminum production from scrap requires less than 10 million Btu per ton.

Similarly for space heating, the efficiency listed in table 12.1 is for heating a building of given physical characteristics, which include the degree of insulation, whether or not there are storm windows, etc. Adding insulation and improving furnace efficiency are complementary approaches to reducing fuel consumption.

Ross and Williams

## Potential Savings

We have described elsewhere in some detail fuel conservation opportunities in four illustrative areas which together account for 40 percent of total U.S. energy use: the automobile, residential space heating, commercial air conditioning, and industrial process steam.[4] In each case the potential savings that could be achieved over the next 10 to 15 years are indicated. Here we only briefly highlight the principal results.

• In the case of the *automobile*, presently available technology could be introduced in the next couple of years to boost average fuel economy in new cars to over 20 miles per gallon, with only a modest reduction in auto weight (say 20 percent). Going further, technological innovations like new engine designs (Diesel, Rankine, or Stirling) and improved transmissions, could lead to an average fuel economy of 30 to 35 miles per gallon for new cars after a decade or so. Of course, these goals have already been achieved with small cars.

• In the area of *residential space heating* modest innovations in design and development of new materials such as better windows and improved insulation to reduce both heat conduction and air infiltration could lead to cutting heat losses in homes nearly fourfold. Such a reduction has far-reaching implications for the heating system, because except for very cold days no supplemental heating beyond what is provided by sunlight through the windows, the electric load, and body heat would be needed. But even further savings possibilities exist for days when modest supplemental heating is needed. A small electric heat pump that uses well water or lake water as a heat source would be twice as efficient in providing heat as a gas furnace.

• It may come as a surprise that in the cost of *commercial air conditioning* the heat generated by lighting is often the largest component of the air conditioning load, accounting for up to 60 percent of the total. Here substantial savings can be achieved by adopting task-specific instead of uniform lighting strategies now often employed, and by turning off lights when they are not in use. In new buildings, greater use of natural lighting could be achieved. After lighting, the next largest component of the air conditioning load is typically the cooling requirement for the ventilation system. While a certain amount of outside air is needed to control odors, to keep carbon dioxide levels down, and to provide adequate oxygen, typical ventilation rates are far in excess of what is required. Moreover, the use of heat exchangers in the ventilation system could substantially reduce the air conditioning requirements for the fresh air that is needed.

With reduced use of lighting, an improved ventilation system, and more insulation, air conditioning demand could typically be reduced to less than one-third its present level. With this greatly reduced air conditioning demand it becomes feasible to think of meeting a substantial fraction of the energy requirements for air conditioning in a commercial office building with solar energy. Solar-assisted, heat-driven air conditioners may well be commercially available within a decade. Implementing all these innovations could cut fuel requirements for air conditioning in a typical New York City office building to one-sixth of what they are now.

• In producing *steam* for industrial process use today, fuel is burned to boil water much as one boils water in a tea kettle. While the first-law efficiency for this process is an impressive 85 percent (that is, 85 percent of the fuel energy ends up in the steam), the second-law efficiency is typically a much more modest 34 percent. The usual process of steam generation wastes the high quality energy in fuel. If instead the combustion energy of the fuel is used first to produce electricity, with the "waste heat" from power generation utilized as process steam, the efficiency of combined electricity and steam production could increase to 40 or 50 percent over the value of 33 percent that characterizes the separate production of steam and electricity.

The resulting savings are actually much more impressive when expressed another way. If only the excess fuel beyond what is required for steam generation is allocated to power production, the fuel required to produce a kilowatt-hour of electricity is reduced to about half of that required in conventional power plants. The national potential is truly great, as process steam is a major energy consuming activity in the economy—accounting for about 15 percent of total energy use.

The most promising method for steam-electricity cogeneration appears to be in industrial plants where electricity is produced as a byproduct, whenever steam is needed. Various cogeneration technologies could be employed. In a steam turbine system, steam would be exhausted from the turbine at the desired pressure and (instead of being condensed with cooling water, as at a conventional power plant) delivered to the appropriate industrial process. With a gas turbine system, the hot gases exhausted from the turbine would be used to raise steam in a waste heat boiler. Of these two systems a major advantage of the gas turbine is that it allows for the generation of much more electricity for a given steam load.

Recent studies on the overall potential for cogeneration have been carried out by Dow Chemical Company[5] and by Thermo Electron Corporation.[6] The Thermo Electron study shows that by 1985 electricity amounting to more than 40 percent of what is consumed in the United States today (generated with the equivalent of about 135,000 megawatts electrical of baseload central station generating capacity) could be produced economically with gas turbines as a by-product of process steam generation at industrial sites. (While the most common gas turbines in use today must be fueled with gaseous or liquid fuels, good experience has been achieved with coal-fired closed cycle gas turbines, and it is likely that by the 1980s relatively low cost technology will be commercially available for firing gas turbines directly with coal.[7])

We turn now to estimating the potential fuel savings from pursuing fuel conservation measures throughout the economy in the United States. Because we wish to focus on what we feel can be implemented on a wide scale within roughly the next 15 years, the proposals we will take into account are somewhat less ambitious than some of those we have discussed so far. The savings for various sections of the economy that would have occurred in 1973, if specific fuel conservation measures had been implemented then, are shown in table 12.2. The potential savings for the economy as a whole are summarized in figure 12.1, where actual and hypothetical energy budgets are compared.

### Zero Energy Growth

It is interesting to note that if the fuel conservation measures considered here had been in effect in 1973, fuel consumption would have been 40 percent less than it actually was. These savings do not take into account what could be additionally achieved through measures involving life-style changes: a heavy shift to mass transit or small cars, enforced 55 mile per hour speed limits, turned down thermostats in the winter, and the like.

One way to interpret the significance of these results is to note that if all the conservation measures considered here had been in effect in 1973, 18 more years of energy growth at the historical rate (3.2 percent per year for the 25-year period ending 1973) would have been required to reach the actual level of 1973 consumption. With an aggressive energy conservation program, it is reasonable to assume that within this time frame the wide use of various fuel saving measures comparable to those indicated in table 12.2 could be achieved. This strongly suggests the possibility of being able to pursue something like *zero energy growth* (ZEG) from now until the

Table 12.2
Projected fuel savings in various sectors of the economy (in quadrillion Btu per year)

| Conservation measures | Potential savings |
|---|---|
| **Residential sector** | |
| Replace resistive heating with heat pump having a COP = 2.5[a] | 0.60 |
| Increase air conditioner efficiency to COP = 3.6[b] | 0.40 |
| Increase refrigerator efficiency 30 percent | 0.27 |
| Cut water heating fuel requirement in half[c] | 1.07 |
| Reduce heat losses 50 percent with better insulation, improved windows, reduced infiltration[d] | 3.30 |
| Reduce air conditioner load by reducing infiltration to 0.2 air exchanges/hour[e] | 0.42 |
| Introduce total energy systems into half of all multi-family units (15 percent of all housing units) with a net 30 percent fuel savings[f] | 0.31 |
| Use microwave ovens for one-half of cooking, with 80 percent savings[g] | 0.25 |
| Total savings | 6.62 |
| Actual fuel use in 1973 | 14.07 |
| Hypothetical fuel use with conservation | 7.45 |
| **Commercial sector** | |
| Increase air conditioner COP 30 percent | 0.37 |
| Increase refrigeration efficiency 30 percent | 0.20 |
| Cut water heating fuel requirements in half | 0.31 |
| Reduce building lighting energy by 50 percent | |
|    Direct savings[h] | (0.82) |
|    Increased heating requirements[i] | (−0.21) |
|    Reduced air conditioner requirements[j] | (0.34) |
|    Net savings | 0.95 |
| Reduce heating requirements 50 percent[k] | 2.25 |
| Reduce air conditioner demand 10 percent with better insulation[l] | 0.08 |
| Reduce air conditioner demand 15 percent by reducing ventilation rate 50 percent (to 0.5 air exchanges per hour) and by using heat recovery apparatus[m] | 0.10 |
| Use total energy systems in one-third of all units, save 30 percent | 0.64 |
| Use microwave ovens for one-half of cooking | 0.06 |
| Total savings | 4.96 |
| Actual fuel use in 1973 | 12.06 |
| Hypothetical fuel use with conservation | 7.10 |
| **Industrial sector** | |
| Good houskeeping measures throughout industry (except for feedstocks)—save 15 percent[n] | 3.85 |

Ross and Williams

Table 12.2 (continued)

| | |
|---|---|
| Fuel instead of electric heat in direct heat applications[o] | 0.17 |
| Steam/electric cogeneration[p] | 2.59 |
| Heat recuperators or regenerators in 50 percent of direct heat applications—save 25 percent[q] | 0.74 |
| Electricity from bottoming cycles in 50 percent of direct heat applications[r] | 0.49 |
| Recycling of aluminum in urban refuse[s] | 0.10 |
| Recycling of iron and steel in urban refuse[t] | 0.11 |
| Fuel from organic wastes in urban refuse[u] | 0.70 |
| Reduced throughput at oil refineries[v] | 0.87 |
| Reduced field and transport losses associated with reduced use of natural gas[w] | 0.80 |
| Total savings | 10.43 |
| Actual fuel use in 1973 | 29.65 |
| Hypothetical fuel use with conservation | 19.22 |
| **Transportation sector** | |
| Improve auto economy 150 percent[x] | 5.89 |
| 35 percent savings in other transportation areas | 3.20 |
| Total savings | 9.09 |
| Actual fuel use in 1973 | 18.96 |
| Hypothetical fuel use with conservation | 9.87 |

Note: If these fuel conservation measures had been in effect in 1973, fuel consumption would have been 40 percent less than it actually was. This suggests that we could pursue zero energy growth from now until the early 1990s without jeopardizing overall economic growth.
[a] APS, *Efficient Energy Use*, sec. 3.C.2. [b] According to Moyers, the best room air conditioning units have a coefficient of performance (COP) twice as large as the present average, 1.8. (See John C. Moyers, "The Room Air Conditioner as an Energy Consumer," Oak Ridge National Laboratory report ORNL–NSF–EP–59, Oct. 1973.) Also some commercial central air conditioning units have a COP = 3.6. We assume the present average is 2.5 for central air conditioning. [c] Current efficiencies are low (APS, *Efficient Energy Use*, sec. 3.D). There are various possibilities for reducing losses: better insulation, reduced hot water heater temperature setting, use of solar energy, or heat recovery from other appliances such as refrigerators. [d] APS, *Efficient Energy Use*, sec. 3.B. [e] APS, *Efficient Energy Use*, sec. 3.C.1. [f] APS, *Efficient Energy Use*, sec. 3.E. [g] Stanford Research Institute, *Patterns of Energy Consumption in the United States* (Stanford, Calif.: The Institute, 1972), p. 46. [h] APS, *Efficient Energy Use*, sec. 3.C.4. [i] Assume that for six winter months all lighting electricity saved must be replaced by fossil fuel heat. [j] According to Irvine, lighting in a typical New York City office building accounts for about 54 percent of the air conditioner load. (See R. Gerald Irvine, testimony in case [no. 26292] before New York Public Service Commission, in *Report on Energy Conservation in Space Conditioning*, Jan. 31, 1974, p. 171.) We assume this is typical. [k] Irvine, *Report on Energy Conserva-*

Table 12.2 (continued)

*tion,* p. 168. [l]Irvine, *Report on Energy Conservation,* pp. 170–171. [m]Irvine, *Report on Energy Conservation,* pp. 171–172. [n]According to Berg and other savings on this order should be possible with better management practices and no changes in capital equipment. (See Charles A. Berg, "Conservation in Industry," *Science* (April 19, 1974), p. 264; and National Bureau of Standards, Institute of Applied Technology, "Energy Conservation Program Guide for Industry and Commerce," NBS Handbook 115 (Washington, D.C.: The Bureau, Sept. 1974). [o]At this point we take no credit for use of recuperators, etc. We assume that only 50 percent of the energy value of the fuel goes to process. (See Berg, "Conservation in Industry.") [p]Here various cogeneration schemes were considered. (See Nydick, et al., "A Study of Inplant Electric Power Generation.") [q]See Berg, "Conservation in Industry," and Gyftopoulos, Lazaridis and Widmer, *Potential for Fuel Effectiveness in Industry* (n. 3). [r]See Elias Gyftopoulos and T. F. Widmer, "Effects of Improved Fuel Utilization on Demand for Fuels and Electricity," in *Air Quality and Stationary Source Emission Control,* a report by the Commission on Natural Resources, NAS-NAE/NRC, prepared for the Committee on Public Works, U.S. Senate, March 1975, chap. 9. [s]Assumes recovery of 0.75 million tons of aluminum from urban refuse (see W. E. Franklin, et al., "Potential Energy Conservation from Recycling Metals in Urban Solid Wastes," in *The Energy Conservation Papers,* R. H. Williams, ed., a report of the Ford Foundation's Energy Policy Project), saving 135 million Btu per ton. This assumes an efficiency improvement in producing primary aluminum. See Gyftopoulos, Lazaridis and Widmer, *Potential for Fuel Effectiveness in Industry.* [t]Assumes recovery of 10.6 million tons of iron and steel in urban refuse (see Franklin et al., in *The Energy Conservation Papers*), saving 10.4 million Btu per ton. This assumes an efficiency improvement in producing steel from virgin raw materials. See Gyftopoulos, Lazaridis and Widmer, *Potential for Fuel Effectiveness in Industry.* [u]Assumes 75 percent recovery of organic wastes and a 70 percent conversion efficiency to a suitable fossil fuel supplement (see Alan Poole, "The Potential for Energy Recovery from Organic Wastes," in *The Energy Conservation Papers*). [v]On the basis of refinery fuel consumption at a rate of 10 percent of output (see Gyftopoulos, Lazaridis and Widmer, *Potential for Fuel Effectiveness in Industry.*) This takes into account fuel conservation opportunities in petroleum refining. [w]Assuming 6.3 percent of gas is consumed in oil and gas fields and 3.3 percent is consumed in pipelines (see U.S. Bureau of Mines, *Minerals Yearbook,* Vol. 1, *Metals, Minerals and Fuels* (Washington, D.C.: The Bureau, 1970), p. 740). [x]APS, *Efficient Energy Use,* sec. 4.G.

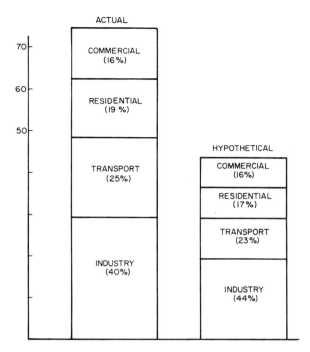

**Figure 12.1**
Hypothetical and actual energy consumption for 1973. The hypothetical budget shows the energy savings potential resulting from investment in conservation equipment based on present technology that is economically justified.

early 1990s without jeopardizing overall economic growth or, in fact, without significantly shifting the course of economic development.

One way of envisioning this is to consider the aggregate demand for products in the economy growing in this period at the historical rate, with the specific energy involved (that is, the energy required to produce a unit of product) being reduced through conservation at a comparable rate, so that the net growth in energy could be small or even zero.

## Conservation for Long Term

Pursuing this program would by no means exhaust the possibilities for fuel savings. Looking beyond 1990, we can get a rough idea of how much opportunity exists for further conservation by estimating the efficiency of

Energy Efficiency

the technologies in place at that time, using the second-law efficiency measure.

Even after pursuing zero energy growth the efficiencies in the residential, commercial, and transportation sectors in the 1990s would still be relatively low—typically in the range 8 to 15 percent, compared to 2 to 10 percent today—and opportunities for substantial further improvements through technological innovation would exist. However, in some parts of the industrial sector at that time operations may well be approaching practical limits to efficiency improvement. In this event, any additional fuel savings may well result from a shift in the industrial product mix: a shift toward less energy intensive products, as might characterize a "post-industrial society." (By less energy intensive, we mean that fewer Btus are required to produce a dollar of output.) Of course, we could adopt policies to encourage a shift in industrial growth even now toward less energy intensive areas. This would make the goals of fuel conservation that much easier to achieve.

### Electrical Energy

Electricity plays a special role in projections of energy use. We have argued that it is desirable to move rapidly to zero energy growth. However, large electric power plants take a long time to plan and construct; many are now under construction. Furthermore, it is desirable in many areas of energy use to continue the trend toward electrification. The electric power capacity under construction now was designed to accommodate historical growth to about 1980. Stretched out to 1985, the capacity now under construction would still allow for growth in electrical energy use from central power plants at about 3.5 percent per year. If this were supplemented with some industrial cogeneration, total growth in electrical energy use would be even faster. Combining this kind of electric power growth for the next decade with something like zero energy growth overall would allow us to approach self-sufficiency in energy by 1985, even with no increases in domestic production of oil and gas.

Beyond 1985, with continued zero energy growth, it would still be desirable to continue the electrification of the energy economy, not only because it will help move us away from dependence on oil and gas but also because electric power generation is relatively efficient, contrary to the popular notion that it is very wasteful. It is only for certain end uses that the second-law efficiencies of electrical energy are often so incredibly low.

But here, as we have said, there are significant opportunities for improvement through technological change.

While it is sensible to continue the trend toward electrification until the year 2000, it would not be sensible to sustain electrical growth nearly as fast as the roughly 4 percent growth rate that would arise in the next decade as a result of finishing the central station plants now under construction, supplemented with some industrial cogeneration. And certainly this nation doesn't need anything like the 6 to 7 percent growth which has been traditionally planned for.

## Economic Benefits

The energy savings from slower or zero energy growth would be reflected in dollar savings from reduced capital costs and from reduced consumer expenditures for energy as well. For some time to come the capital investment required to save a kilowatt of power is likely to be much less than the capital investment required to generate a kilowatt from a new energy source. A serious commitment to fuel conservation would require adoption of policies that would make available to fuel conservation projects some of the capital freed by not having to build as many new energy generation facilities. Although using energy more efficiently would save the consumer money in the long run, increased first costs would often be required to realize these savings.

For example, an investment in more insulation to reduce heating fuel consumption might be justified economically. But the individual consumer's decision-making costs and lack of access to extra capital at reasonable interest rates may inhibit such investments. The energy industry may find it easier to make a corresponding investment in energy supply. Tax incentives or government guaranteed loans would probably be required to help finance increased energy efficiency.

Fuel conservation has even broader economic benefits than these considerations suggest. Consider the role of energy in the economy generally. In the past when energy was cheap and abundant it was indeed appropriate to emphasize increased energy inputs to economic activity. The resulting productivity gains led to economic growth and new jobs. But, with the expectation of continued high and perhaps even higher energy prices in the future, the prospects for spurring economic growth with increasing energy use are much diminished.

Contrary to the popular but erroneous view, the alternative—fuel conservation—does not imply economic stagnation. Stagnation and especially dislocation occur, as in 1975, due to sudden changes in the economy. If achieved gradually, a zero energy growth society could have a vigorous economy. A determined national commitment to fuel conservation would create many new job opportunities. New businesses and industries would be needed to produce, market, install, maintain and repair energy conservation technology: new building insulation materials, heat pumps, electronic controls for regulating energy use in buildings, new types of batteries and other local energy storage systems, new coal burning cogeneration devices, retrofit equipment for large air conditioning systems, communications systems that substitute for transportation, and so on.

The small scale of most energy conservation technologies would probably spur productivity gains in the economy as a whole. Efforts to save energy in space heating, for example, could create a demand for millions of heat pumps. Our economy could do what it has done so successfully in the past: develop improved technology through competitive trials, and cut production costs through use of mass processes. In contrast, very large energy supply devices and systems such as characterize nuclear power, do not enjoy these advantages. This may explain the galloping inflation plaguing these systems.

The potential economic impact of the conservation option is illustrated in figure 12.2, where two alternative energy futures are illustrated: one representing a continuation of historical trends; the other a course of just 1 percent average energy growth between now and the year 2000. These paths, simulated with an econometric model developed for the Ford Foundation's Energy Policy Project, involve different assumptions about energy prices.[8] The lower energy growth course results from more pessimistic assumptions about the prices of petroleum products and electricity (for example it assumes that the real price of electricity grows at 1.7 percent instead of −2.7 percent per year), coupled with the imposition of a modest tax on the sales of energy. The model suggests that the cumulative effect by the year 2000 of a gradual shift to greater energy efficiency (in response to higher prices in this model) would be a slight reduction in gross national product, but an increase in employment.

A shift in our nation's course to low and probably zero energy growth may well be forced upon the United States in the not too distant future through damaging confrontations with other energy hungry nations or with environmental limits. Another, perhaps more likely, possibility is that energy growth will be halted by normal economic forces. Unfortunately, a

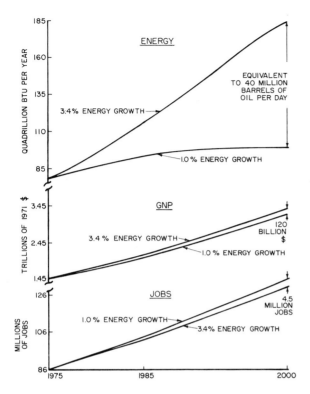

**Figure 12.2**
Alternate future energy courses for the United States. An econometric model of the U.S. economy, assuming two different price structures for energy and two different growth rates—3.4 percent (the historic rate) and 1 percent—between now and the year 2000, predicts that the cumulative effect of a gradual shift to greater energy efficiency would be a slight reduction in GNP and an increase in employment. See Hudson and Jorgenson, "Economic Analysis of Alternative Energy Growth Patterns" (n. 8).

Energy Efficiency

policy of massive subsidization of the energy supply industry, in order to continue growth for the time being, might well lead to an end of energy growth with an economy unable to adjust to efficient energy use without severe dislocation. There is time to cushion the impact of the transition to zero energy growth if the nation begins now to make a serious commitment to efficient energy use. An attractive bonus of this course would be an invigorated economy with vast opportunities for technological innovation.

## Notes

1. Energy Policy Project of the Ford Foundation, *A Time to Choose: America's Energy Future* (Cambridge, Mass.: Ballinger, 1975); and American Physical Society, "Efficient Use of Energy: A Physical Perspective," a report prepared by W. Carnahan et al., in *Efficient Energy Use, American Institute of Physics Conference Proceedings*, August 1975, vol. 25 (New York: The Society, 1975). A summary report of the APS study appears in *Physics Today*, 28:8 (Aug. 1975), 23–33.

2. Raymond W. Bliss, "Why Not Just Build the House Right in the First Place," *Bulletin*, 32 (March 1976), 32.

3. The theoretical basis for doing second-law efficiency calculations was provided in the pioneering work of physicist Willard Gibbs nearly 100 years ago. Unfortunately, the diffusion time for this science into public policy applications has been long. Over the past several years, Charles Berg ("A Technical Basis for Energy Conservation," *Technology Review* (Feb. 1974), p. 14,) and the Federal Power Commission staff report, FPC/OCE/2 (April 1974), has advocated use of this concept in public policy-making. A study done for the Energy Policy Project by E. P. Gyftopoulos, J. J. Lazaridis, and T. F. Widmer (*Potential for Fuel Effectiveness in Industry* [Cambridge, Mass.: Ballinger, 1974]) applied the concept to industrial energy use, and the study by the American Physical Society (n. 1) estimated second-law efficiencies for energy consuming activities throughout the economy.

4. M. H. Ross and R. H. Williams, "Assessing the Potential for Fuel Conservation," *Technology Review*, (February 1977), 49.

5. "Energy Industrial Center Study," report of Dow Chemical Company, the Environmental Research Institute of Michigan. Townsend-Greenspan and Company, and Cravath, Swaine, and Moore to the National Science Foundation, June, 1975.

6. S. E. Nydick et al., "A Study of Inplant Electric Power Generation in Chemical, Petroleum Refining, and Paper and Pulp Industries," draft report prepared by Thermo Electron Corporation, Waltham, Mass., for Federal Energy Administration, May 1976.

7. In West Germany a number of small-scale coal-fired closed cycle gas turbines producing both electricity and heat have been successfully operated for more than 10 years. See K. Bammert and G. Groschup, "Status Report on Closed Cycle Power Plants in the Federal Republic of Germany," paper presented at Gas Turbine Conference and Products Show of American Society of Mechanical Engineers, New Orleans, March 21–25, 1976, paper no. 76–GT–54, *Transactions of American Society of Mechanical Engineers*.

Perhaps the most promising technology for the future is high pressure fluidized bed combustion of coal. For a discussion of this technology used for cogeneration, see R. H. Williams, "Industrial Cogeneration," Center for Environmental Studies Report No. 66 (Princeton, N.J.: The Center, May 1978).

8. Edward A. Hudson and Dale W. Jorgenson, "Economic Analysis of Alternative Energy Growth Patterns, 1975–2000," in *A Time to Choose*, Appendix F.

The flow of solar energy through the earth's natural systems is about ten thousand times greater than the flow of fossil fuel energy into man's machines. Indeed, for most purposes, fossil fuel energy provides only a small supplement to products that are already supplied primarily by natural processes driven by solar energy: light, warmth, organic materials (such as food, wood, paper, fibers, etc.), clean air, and water, for example. It seems reasonable therefore that man could in one way or another use his ingenuity to rechannel a small amount of the flow of solar energy to reduce his dependence on fossil fuels.

As it appears that petroleum and natural gas can supply a major fraction of the energy needed to support the technological superstructure of our civilization for only a few more decades, alternative energy supply sources must be developed for the long term. The principal candidates besides solar energy are coal, nuclear fission, and controlled thermonuclear fusion.

The conventional wisdom is that, for the foreseeable future, man will have to shift increasingly to coal and fission to replace dwindling supplies of petroleum and natural gas. Fusion is still an unproven technology for which commercialization on a large scale is probably at least 50 years off. Many solar technologies are proven. But most members of the energy-supply "establishment" believe (almost reflexively) that solar energy is, and will remain, too expensive to collect and store for the bulk of our energy requirements.

There are fundamental problems associated with primary dependence on coal or fission energy for the long term, however. We believe, therefore, that a very high priority should be given to an assessment of whether it would be feasible to establish a viable energy economy based primarily on solar energy within the next half century.

We do not know, as yet, how far we can shift toward an "all" solar energy economy. There may be fundamental technological, economic, and institutional constraints that will limit the degree to which a shift toward a solar energy economy is feasible in a society dependent, as heavily as ours is, on energy intensive technologies. We believe, however, that only by considering the possibility of making solar energy the principal replacement for fossil fuels in the next century can the associated issues be directly confronted. An understanding of these issues is important both for the setting of solar energy research and development priorities and for the formulation of energy policy.

# 13   Toward a Solar Civilization
# Frank von Hippel and
# Robert H. Williams

## The Alternatives

*Coal* Estimated world resources of coal could support human fossil fuel use at the present level for about 500 years. It is quite possible, however, that well before 20 percent of this resource is consumed, the environmental and social impact of coal burning would prove to be intolerable.

We are not referring just to the problems of air pollution, land destruction, and occupational health. Although these are serious problems, they can, in principle at least, be mitigated to a substantial degree with technological controls. Our primary concern is the cumulative impact of coal burning on the global climate—that is, the impact from the buildup of the level of carbon dioxide in the earth's atmosphere as a result of the worldwide consumption of fossil fuels.

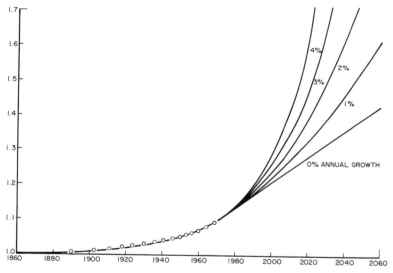

**Figure 13.1**
Buildup of carbon dioxide in earth's atmosphere. Historical data and projected concentration of carbon dioxide in the atmosphere (pre-industrial level = 1.0) for different assumed growth rates of fossil fuel combustion. The curves are based on the assumption that 50 percent of the carbon dioxide released stays in the atmosphere. Note that, even at current rates of fossil fuel consumption, the excess carbon dioxide concentration due to a man's activities would approximately double between 1970 and 2000.

Von Hippel and Williams

This buildup of carbon dioxide in the atmosphere is already so rapid (see figure 13.1), and the period over which a shift to a predominant dependence on nonfossil fuel energy technologies might realistically occur is so long (perhaps 50 years), that this problem should be a major consideration in current energy planning.

The primary concern about the buildup of carbon dioxide in the atmosphere is that it will lead to global warming—the "greenhouse effect." A recent review of the literature concludes that—with a 2 percent annual rate of increase in worldwide consumption of fossil fuels (that is, essentially zero growth in per capita use of fossil fuels at the current world population growth rate)—the average global temperature would increase by 0.5 to 2.5°C from the carbon dioxide buildup in the next 50 years.[1]

Although such changes in the average global temperature may appear to be relatively small, the impact of the associated shifts of weather patterns could be quite dramatic,[2] though mixed. Some currently semiarid regions would probably experience increased rainfall and some northern areas would probably enjoy longer growing seasons. But some important food producing regions could be seriously damaged.[3]

It is difficult to predict who would gain and who would lose from these shifts. As global climate modeling is so difficult, and the atmospheric level of carbon dioxide is rising so fast, we might discover the consequences only by experience. In such an uncertain situation, it seems imprudent to rely on coal as our principal energy option for the coming century.

*Nuclear Fission* With uranium conserving reactors, such as the plutonium breeder reactor, nuclear fission could support U.S. energy use at its current level for thousands of years. It is questionable, however, whether nuclear fission can be exploited worldwide on a large scale over the long term without significantly increasing the likelihood of nuclear violence.

Indeed, with the global deployment of a plutonium economy (the option favored by most nuclear power establishments in the world today) maintaining the separation of the peaceful and the weapons atom could become totally impossible. Although some uranium conserving fuel cycles may be more proliferation resistant than the plutonium fuel cycle,[4] such systems would still require rigid international controls which might not be achievable nor maintainable over the long term.

These then are the principal reasons why we believe that it is urgent to explore a solar energy economy.

## A Solar Strategy

Solar energy is diffuse, intermittent, ubiquitous, and diverse in its manifestations. These properties imply that a well-designed energy economy based on solar energy would be far different from the energy economy we know today. This fact does not seem to have been taken into account in the present debate over solar energy. Indeed, a shift to a solar economy might reshape our way of life at least as profoundly as did the introduction of the automobile or the products of the electronics industry.

The properties of sunlight imply that a solar system will generally require larger equipment to provide a kilowatt than a conventional energy system, so that innovative strategies will be required to keep capital costs reasonable. Large collection areas are required for solar systems (large, for example, relative to the areas of the heat exchangers which collect comparable amounts of energy in fossil and fission fuel powered plants), because solar energy is diffuse. And bulky storage would often be required to provide energy when the solar resource is not available. (For example, it would take 100 gallons of water heated from $20°C$ to $100°C$ to store an amount of energy equal to the chemical energy stored on one gallon of gasoline.)

Some of the changes needed if a solar energy future is to be viable will come about whatever energy supply course is pursued. One of these is improved efficiency in the use of energy. As a result of the energy price increases which have already occurred and which can be expected in the future, there are many potential energy efficiency improvements today which will represent a better use of capital than any new energy supply technology.

A shift to a solar energy-based economy would involve more far-reaching changes in our energy supply-utilization system, however.

## Sharing Capital Costs

Since solar energy is ubiquitous, there are many potential synergisms between human activities and their solar energy environment that could be exploited to reduce capital costs.

In a house, which is thermally tight to begin with and which has a southern exposure, for example, most of the winter space heating requirements could be satisfied by solar energy. Large, south facing windows with overhangs could let in the low angle winter sunlight, but keep out the high angle summer sunlight.[5] Heavy interior walls and floors of masonry or concrete could act as thermal "shock absorbers" by soaking up excess midday solar heat and releasing it later at night and during cloudy spells. Only the extra costs beyond those which would be associated with nonsolar construction would have to be charged to the solar heating system.

Another way by which the costs of solar energy conversion systems could be shared with other activities would be to site solar energy conversion units in locations where by-products having economic value could be utilized.

One example would involve the decentralization of electricity generation to a level where the reject heat could be used locally for industrial process heat (via cogeneration systems) or for residential or commercial space conditioning (via total energy systems).[6]

It is interesting to note in this connection that one recent review of solar electric power technologies, with the subtitle "solar electricity can make only a limited contribution to the nation's large-scale energy needs," neglected entirely the possibility of solar cogeneration or total energy facilities.[7] This may reflect the degree to which large-scale central station power plants have come to dominate the thinking of the electricity supply community.

Cost cutting combinations can involve nonenergy by-products as well. For example, biogasification (anaerobic digestion by bacterial action) of agricultural wastes (manure, corn stalks, etc.) yields in addition to methane gas a residual sludge with fertilizer value.[8]

### Appropriate Scale

Coal or uranium-fueled power plants can be constructed so compactly that single units are being built with generating capacities exceeding 1,000 megawatts (electric). In contrast, because of the diffuseness of solar energy, it seems unlikely that solar electric power plants, with their large collector areas, would benefit from economies of scale much beyond about 10 megawatts average power. This corresponds to about 100 acres (40 hectares) of land area devoted to the collector for a solar power plant operating at 15 percent overall conversion efficiency.

The ubiquity of the solar energy resource and the difficulties associated with long distance transmission and distribution of the reject heat energy in cogeneration or total energy configurations are other factors that suggest solar electricity generation would be most economic on a scale where an individual facility supplied electricity to only hundreds or thousands instead of hundreds of thousands or millions of individuals.

While there are economic disincentives to building very large scale solar energy systems, there might also be substantial economic penalties associated with deploying systems which are too small.

For example, while the heat storage capacity of a hot water tank or a bin of hot rocks increases with the storage volume, the storage costs tend to increase with the surface area—between a certain minimum size and a maximum size where some dimension (usually depth) approaches its practical limit. Opportunities for realizing scale economies in this range of storage capacities suggest that long-term storage of heat (or coolth for air conditioning) would best be carried out at the community rather than at the household level.

Community rather than household scale solar collection systems would also be favored if collectors which "track" the sun and concentrate its radiation prove to be more economical than stationary flat plate collectors. While flat plate collectors tend to be bulky, tracking collectors could have lightweight reflective surfaces which would concentrate the sun's energy on a relatively small boiler or other energy converter. (One very lightweight tracking collector, sheltered from the weather in a transparent air-supported bubble, is shown in figure 13.2)

It would probably be worthwhile to exploit the concentrated energy collected at the receiver of such a system for the combined generation of electricity and heat in total energy systems. Such relatively complex arrangements would probably be more practical at the community than at the household level.

Collection and conversion of solar energy on a community scale would also leave the design and siting of individual houses unconstrained by massive solar hardware.

### Regional Strategies

Unlike conventional energy technologies there would probably not be any single "best way" to exploit solar energy all over the world or even around the United States. Today's highly integrated petroleum and natural gas networks, which supply primary energy on a national and international scale using large tankers and pipelines, have brought about a great similarity between the energy conversion systems in different regions and nations.

**Figure 13.2**
Sun-tracking mirror collector. This mirror, enclosed in an air supported plastic bubble, is part of a solar energy collecting system being developed by the Boeing Company. The sun's rays reflected from an array of such "sun-tracking" mirrors would be concentrated on a central receiver, where high temperature solar heat would be recovered. The purpose of the bubble is to protect the mirror from the weather. As a result, the mirror can be lightweight and, hopefully, low cost.

(They have also brought with them a vulnerability to disruptions on a global scale, such as was caused by the 1973 oil embargo.)

In contrast, most collected forms of solar energy are not easily transported over long distances, and different regions are endowed with quite different solar energy resources and mixes of energy demands. For example, in the United States,

● the direct collection of solar energy is favored in the sunny Southwest,
● the production of chemical fuels from agricultural wastes in the corn belt of the Midwest,
● the deployment of windmills on the Great Plains, and
● the siting of ocean thermal gradient electricity generators off the coast of the Gulf of Mexico. (An ocean thermal gradient system would produce electricity by extracting useable heat from warm surface waters and reject-

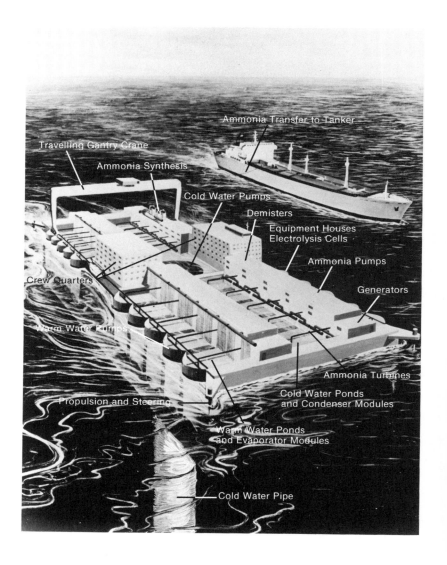

Ammonia Transfer to Tanker

Travelling Gantry Crane

Ammonia Synthesis

Cold Water Pumps

Demisters

Equipment Houses
Electrolysis Cells

Ammonia Pumps

Crew Quarters

Generators

Warm Water Pumps

Ammonia Turbines

Propulsion and Steering

Cold Water Ponds
and Condenser Modules

Warm Water Ponds
and Evaporator Modules

Cold Water Pipe

Von Hippel and Williams

ing the waste heat from power generation to cold waters pumped up from deeper layers of the ocean.[9])

If large-scale generation of electricity in remote locations turns out to be economically advantageous with technologies such as offshore wind energy conversion plants or floating ocean thermal gradient electrical generators which "graze" in tropical waters, it might make sense to minimize the associated energy transmission problems by locating at these sites major energy consuming facilities such as aluminum or ammonia plants (see figure 13.3). Such developments have already occurred in the case of the aluminum industry, which has been attracted to remote sites by a particularly cheap form of solar energy: hydropower.

## Matching End Uses

Some forms of solar energy (such as direct sunlight, wind or biomass) are more naturally converted into some intermediate energy forms than others. The extent to which each solar energy resource would be exploited in a solar based economy would therefore depend upon the mix of energy forms required.

A breakdown of primary energy requirements by end use for the U.S. economy in 1973 is presented in table 13.1. While the pattern of energy uses will, of course, change over time, the breakdown in the table is suitable for the purpose of an initial discussion.

**Figure 13.3**
Ammonia-producing factory grazing tropical ocean waters. Artist's conception of an ammonia-producing factory ship grazing tropical ocean waters. The temperature difference between the warm surface waters and cold deep ocean waters drives heat engines that power electrical generators. The electric power is used to produce hydrogen (by the electrolysis of water), which is then combined under pressure at about 400° C with nitrogen extracted from the atmosphere to make ammonia ($NH_3$). Most of the volume of the ship is occupied by warm and cold water ponds (along the sides and center-line respectively). The warm water ponds are continually replenished through the short pipes hanging over the sides of the ship while the cold water pond is replenished from the deep ocean through a 60-foot diameter, one-half mile long pipe. The working fluid of the heat engine first vaporizes in heat exchangers in the warm water ponds. It then expands through turbines which drive the electric power generators. Finally it condenses again by passing through heat exchangers in the cold water ponds. Source: Evan J. Francis and Joseph Seelinger, "Forecast Market and Shipbuilding Program for OTEC/Industrial Plant-Ships in Tropical Oceans," *Proceedings of 1977 Annual Meeting of American Section of the International Solar Energy Society*, pp. 24–28.

Table 13.1
Distribution of U.S. primary energy consumption by end use in 1973

| End Use | Percentage of total primary energy use | |
|---|---|---|
| Water heating | 4 | |
| Space conditioning | 19 | |
| Industrial process heat | 24 | |
| Low temperature (process steam) | | 14 |
| High temperature (direct heat) | | 10 |
| Cooking and clothes drying | 1.5 | |
| Transportation | 25 | |
| Urban auto use and small trucks | | 11 |
| Rail | | 1 |
| Rural auto use and large trucks | | 6.5 |
| Air | | 2.5 |
| Other | | 4 |
| Miscellaneous electric | 19.5 | |
| Electric drive | | 9 |
| Refrigeration | | 3 |
| Lighting | | 5 |
| Electrolytic processes | | 1 |
| Other | | 1.5 |
| Feedstocks and other nonfuel uses | 7 | |
| Total | 100 | |

Source: M. H. Ross and R. H. Williams, "The Potential for Fuel Conservation" (Albany: Institute for Public Policy Alternatives, State University of New York, July 1975).

About 35 percent of the U.S. energy demand in 1973 was for low temperature heat (including hot water, space heat, and industrial process steam); about 20 percent for end uses which are currently and probably will continue to be electric ("miscellaneous electric"); and about 20 percent for end uses which appear to be best matched with chemical fuels (long distance transport and non-fuel uses). The remaining activities (mainly railroads, urban transport, and high temperature industrial process heat) could be supplied with some mix of electricity and chemical fuels. (Solar "furnaces"—accurately, sun tracking collectors with concentration ratios of 1,000 or more—could be used as an alternative source of high temperature industrial process heat.)

Von Hippel and Williams

In what follows we discuss strategies for producing low temperature heat, electricity and chemical fuels with solar energy.

*Low Temperature Solar Heat* Solar water heating is already being extensively utilized in some parts of the globe. Millions of rather rudimentary solar hot water heaters are used in Japan, and solar hot water heaters with insulated storage tanks are widely used in Israel and Australia. In the United States, before the availability of cheap natural gas, tens of thousands of solar water heaters were in use in Florida.[10] Now that natural gas and electricity are becoming more expensive, solar water heaters are being installed in considerable numbers throughout the United States.

Conditions favoring the relatively widespread use of solar water heating in the United States are as follows:

• The collectors can be very simple and relatively small. They can therefore benefit from the economies of mass production. It is also relatively easy to find space to install them.

• Hot water is required throughout the year, so that relatively little collectable solar energy need be wasted.

• The required energy storage system is somewhat greater in volume but not different in kind from what is used with conventional water heaters.

Solar space heating is in an earlier stage of deployment (see figure 13.4). Because the collecting areas and storage volumes required are more comparable to the areas and volumes of the buildings being heated, the integration of the solar space heating system into the building design requires more thought than a water heating system. Thus far no clear consensus has developed as to which approaches will ultimately be preferred.[11]

For solar space heating the cost of the collected energy is driven up by the fact that the collectors are not used during the warmer parts of the year. This has led to considerable interest in "annual cycle energy storage" systems in which heat collected in summer is stored for the winter.[12]

An alternative use of the excess summer heat from solar collectors is for heat driven air conditioning cycles. Indeed, heat driven air conditioning may be of interest even in areas where the air conditioning demand is much greater than the space heating demand. This is partly because air conditioning loads tend to peak when direct solar heat is available, thereby reducing the need for storage for a solar driven system.

Peak load pricing for electricity (increasingly being urged in such areas)

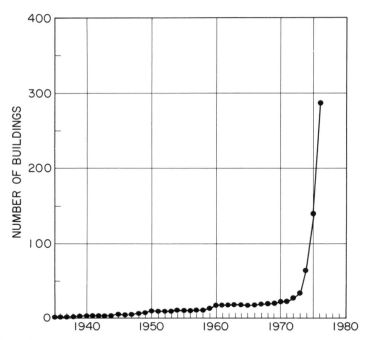

**Figure 13.4**
Cumulative number of solar buildings completed in the United States (1940–1976).
Source: W. A. Shurcliff, *Solar Heated Buildings: A Brief Survey*, 13th and Final
Edition (19 Appleton St., Cambridge, Mass. 02138), 1977.

would also improve the economic position of solar air conditioning relative
to conventional electric air conditioning.

Chemical plants and paper mills are examples of large-scale users of low
temperature heat in the form of process steam. The scale and the high load
factor (arising from 24-hour, year-round operations) for many process
steam-using activities strongly suggest providing these heating needs with
the reject heat from on-site solar electrical cogeneration systems.

*Solar Electricity* The trend in the U.S. energy economy has been toward
increasing electrification. Propelled by a long-term rapid decline in the real
price of electricity, the demand for electricity has grown over past decades
at twice the rate for total energy. Primary energy consumption for electri-

Von Hippel and Williams

city production accounted in 1950 for about 16 percent of total U.S. energy use. This rose to 29 percent in 1975 and is often projected by the electric utilities as climbing to 50 percent by the year 2000.

It seems unlikely that the historical trend toward electrification will continue at past rates, however. The historical downward trend of electricity prices (in deflated dollars) has reversed itself in the past five years, and it is very likely that the new upward trend will continue for decades. It is unlikely, therefore, that many of the relatively inefficient uses of electricity (such as large-scale resistive heating) which were projected by the Energy Research and Development Administration and the utilities in the past will ever become economic.[13]

The solar electric technologies that have received the lion's share of research and development funds to date are central station technologies. But, as we have already noted, such purely electric systems may not be as economic as smaller scale cogeneration or total energy systems producing both electricity and locally usable heat.

While small scale power generating systems cannot benefit from scale, they can enjoy the compensating economic benefits of mass production. A 500-kilowatt diesel generator set, for example, if built on a "one-of-a-kind" basis would be much more expensive per unit generating capacity than a large modern oil fired steam turbine driven power plant with 1,000 times the power. Yet, once the diesel goes into mass production, the economies of the assembly line match the "economies of scale" and the two systems cost about the same per kilowatt of generating capacity.[14]

It is sometimes suggested that the use of decentralized electricity generating technologies might result in large savings in transmission and distribution costs, which currently account for about half of the cost of electricity to the average consumer.[15] It is not at all clear, however, that the high costs of transmission and distribution would not be offset by the benefits of interconnection, including a sharing of backup capacity among the many dispersed generators, and perhaps also by a reduction in the total electrical storage requirements. The tradeoffs between strong and weak electrical interconnections should be carefully examined.

*Chemical Fuels from Solar Energy* Chemical fuels are a compact medium for long-term storage and long-distance transport of solar energy. In a 100 percent solar economy, chemical fuels such as methane, ethanol, or char oil would probably be needed for at least non-fuel uses and long-distance transport (currently about 20 percent of total energy use).

Chemical fuels may also be desirable for providing high temperature industrial process heat and, in some instances, chemical backup fuel may be less costly than long-term storage of solar electricity or heat.

Two alternative approaches for producing chemical fuels which have been considered are

● the "cracking" of water to produce hydrogen, and
● the conversion of photosynthetically derived organic materials to convenient fuel forms.

Two new technologies which have been proposed for the production of hydrogen from water are the use of heat-driven chemical cycles (where all of the chemicals involved other than water would be regenerated) and photolysis (where the energy of sunlight is captured in photovoltaic materials), but so far neither of these approaches has advanced beyond the research stage.[16]

Electrolysis of water to produce hydrogen is a well-established industrial process. However, it would probably not be economic as a means of deriving large quantities of chemical fuel except under special conditions, such as at remote locations where electricity can be generated relatively cheaply (for example, in regions where there is abundant hydropower or wind power produced with especially high winds), and where there is insufficient demand for the electricity within a practical range for direct transmission.

Perhaps the most economically promising approach for producing limited amounts of chemical fuel is via the *conversion of photosynthetically derived organic wastes.* The energy content of concentrated sources of urban refuse, pulp waste, wood waste in forests, crop residues, and manure from animals in feedlots amounts to about 17 percent of current U.S. energy use.[17] Losses in the conversion of these raw materials to conventional fuel forms, such as methane via biogasification or gas or char oil via pyrolysis would reduce the fuel value of these organic waste resources to about 10 percent of current total U.S. energy use. If the organic materials were reacted with water in a solar furnace to produce hydrogen, however, the fuel produced could actually contain more solar energy than the original organic wastes.[18] (Antal has shown that it should be easier to produce gaseous fuels via steam pyrolysis from organic wastes rather than from coal.[19])

Another way to extend the photosynthetically based chemical fuel resource would be to grow plants as fuel. For selected and carefully managed

Von Hippel and Williams

energy crops, relatively high photosynthetic efficiencies are possible. Recent NASA research suggests, for example, that—if the water hyacinth (which today clogs rivers, ditches, and canals in the Southeast United States) is recovered as an energy resource in sewage treatment ponds (from which the plants extract concentrated nutrients)—the average annual efficiency of conversion of solar to chemical energy could be as high as 5 percent.[20] This is a ten-fold greater efficiency than that of typical natural and agricultural systems in the United States.

Of course, large tracts of land would be required to grow significant quantities of biomass fuels. One would not wish to displace agriculture or forestry with fuel farms, because food and timber products are inherently more valuable per pound than fuel. A land resource that could potentially be used for fuel farming, however, is the roughly 5 percent of U.S. land mass which is periodically waterlogged (fuel farming would require large quantities of water) and is not suitable for ordinary agricultural or forestry uses. Even at only 0.5 percent photosynthetic efficiency (which is typical of natural ecosystems) these lands could yield—via conventional biogasification and char oil production processes—fuels equivalent to more than 10 percent of current U.S. energy consumption.[21]

In brief, at the current level of technology the potential for the economical production of chemical fuels in a solar economy is limited, though perhaps adequate to meet minimum needs for chemical fuels in a solar energy economy where total energy consumption is comparable to U.S. energy use today. If the level of U.S. energy demand should increase very much, however, the chemical fuel shortage in an "all" solar energy based economy might become quite severe.

## A Technological Revolution?

In the era of cheap fossil fuels, energy supply and energy consumption technologies came to be the subjects of different technological cadres who were able to ignore each other almost completely. The design of a house, for example, was not substantially influenced by the nature of the home heating plant except through the distribution conduits. Similarly, electric energy generation became separated from consumers, sometimes by distances of hundreds of miles. In the new era of more expensive energy our architects and engineers, our construction and contracting organizations, and our regulatory and taxing bodies will all have to adopt a more integrated approach to energy supply and use.

As we have indicated, this reorganization of our society will be particularly far-reaching if solar energy comes to be the principal successor of fossil fuels. The difficulties which will be encountered in bringing about such changes should not be underestimated. But they are essential if the world is to avoid a true energy crisis in the decades to come.

The changes will come much more easily if, as appears possible, a broad-based political commitment develops in our society to the idea of shifting toward a major dependence on solar energy. Indeed, society tends to re-arrange itself whenever a desirable new technology "catches on." This adaptability of society was certainly apparent in the technological revolutions that brought us the railroads, the automobile, and the computer. History demonstrates therefore that where there is a will there is (often) a way.

In the case of solar energy, however, the fundamental question is whether there will be adequate incentives for the changes to be made. The railroads, automobiles, and electronics are technologies which created opportunities to do things which were not possible before. In contrast, solar energy will provide just another way to boil water, generate electricity, or produce chemical fuels. In order for society to make the shift to a solar energy economy, therefore, a political consensus may have to develop to the effect that solar energy should be chosen over fission and fossil fuels because it is environmentally and socially more benign—or solar energy will have to be cheaper.

It is not that clear what the long-term result would be from a free economic competition among solar energy, coal and nuclear fission. Coal and fission offer the advantage of concentrated energy supply, so that energy "collection" is relatively inexpensive. Solar technologies have the advantage of simplicity.

Many billions of dollars are required to bring a single new nuclear reactor to the prototype stage. Thousands of man-years are devoted to analyzing just its safety. The construction process is also made enormously complex and bureaucratic in an attempt to assure that safety is not compromised by failure of "quality assurance." In such a world the creative engineer must usually give way to the organization man. The systems are far from the optimum which the technology allows, and no one understands in the end how the plants got to be so complex and expensive.

The relatively small scale of solar technologies gives designers a freedom to experiment and utilize their talents more fully. This has already led to a creative ferment and a rapidly growing community of self-selected solar technologists. Daily we hear of new breakthroughs in the direction of

Von Hippel and Williams

simplicity and low cost in solar systems and their optimization for different regions or end uses.

As with electronics, which is also a basically small-scale technology, it appears to be impossible for any individual or even government to maintain an overview of the research "frontier."

The best approach may be to "let a thousand flowers bloom," and then to cultivate the most promising varieties. We have a society where hundreds of thousands of citizens are skilled in science and technology. It may take a solar technological revolution to remind us once again that the true strength of such a society is only revealed when its members are given the opportunity to show what they can do.

## Notes

1. C. F. Baes, et al., *The Global Carbon Dioxide Problem* (Oak Ridge, Tenn.: Oak Ridge National Laboratory, 1976).

2. Stephen H. Schneider, *The Genesis Strategy: Climate and Global Survival* (New York: Plenum, 1976).

3. William W. Kellogg, "Global Influences of Mankind on the Climate," in *Climate Change*, John Gribbon, ed. (Cambridge University Press, 1977).

4. Harold A. Feiveson and Theodore B. Taylor, "Security Implications of Alternative Fission Futures," *Bulletin*, Dec. 1976.

5. Raymond Bliss, "Why Not Just Build the House Right in the First Place," *Bulletin*, March 1976.

6. See, for example, B. W. Marshall, "Analysis of a 1000-Home Solar Total Energy Community Using Clear Air Solar Intensity," SAND 75-0097 (Albuquerque, N. Mex.) Sandia Laboratories Energy Report, May 1975; and R. L. Pons and R. J. Fox, "A Solar/Stirling Total Energy System," in *Sharing the Sun: Solar Technology in the Seventies*, Joint Conference of the American Section, International Solar Energy Society, and Solar Energy Society of Canada, Inc., K. W. Böer, ed. (Winnipeg: The Society, 1976), 5, 15–20.

7. William G. Pollard, "The Long-Range Prospects for Solar Energy," *American Scientist*, July-August, 1976, p. 424.

8. Alan D. Poole and Robert H. Williams, "Flower Power: Prospects for Photosynthetic Energy," *Bulletin*, May 1976.

9. Clarence Zener, "Solar Sea Power," *Bulletin*, Jan. 1976.

10. Richard Schoen, et al., *New Energy Technologies for Buildings*, a report to the Energy Policy Project of the Ford Foundation (Cambridge, Mass.: Ballinger, 1975).
   Between 1945 and 1960 some 50,000 solar hot water heaters were installed in southern Florida before the availability of cheap natural gas resulted in their being phased out.

11. W. A. Schurcliff, "Active Type Solar Heating Systems for Houses: A Technology in Ferment," *Bulletin*, Feb. 1976.

12. See, for example, *Proceedings of the 1977 Annual Meeting of the American Section of the International Solar Energy Society*, Sec. 16.

13. von Hippel and Williams, "Energy Waste and Nuclear Power Growth," *Bulletin*, Dec. 1976.

14. S. E. Nydick, et al., *A Study of Inplant Electric Power Generation in the Chemical, Petroleum Refining, and Paper and Pulp Industries*, report prepared by Thermo Electron Corporation for the Federal Energy Administration, June, 1976.

15. M. L. Baughman and D. J. Bottaro, "Electric Power Transmission and Distribution Systems: Costs and their Allocation" (Austin, Texas: Center for Energy Studies, University of Texas at Austin, July 1975).

16. R. H. Wentorf, Jr., and R. E. Hannerman, "Thermochemical Hydrogen Generation," *Science*, 185 (July 26, 1974), 311; and Mark S. Wrighton, "The Chemical Conversion of Sunlight," *Technology Review*, May 1977.

17. Poole and Williams, "Flower Power."

18. M. J. Antal, Jr., "Tower Power: Producing Fuels from Solar Energy," *Bulletin*, May 1976.

19. M. J. Antal, "A Comparison of Coal and Biomass as Feedstocks for Synthetic Fuel Production," paper presented at the Alternative Energy Source Symposium, Miami Beach, Florida, Dec. 5-7, 1977.

20. B. Wolverton and R. C. McDonald, "Don't Waste the Waterweeds," *New Scientist*, Aug. 12, 1976, p. 318.

21. Poole and Williams, "Flower Power."

Michael J. Antal, Jr.
Department of Aerospace and Mechanical Sciences
School of Engineering/Applied Science
Princeton University
Princeton, NJ 08540

Raymond W. Bliss
Donovan and Bliss
Chocorua, NH 03817

Richard S. Caputo
SERI (Solar Energy Research Institute)
1536 Cole Blvd.
Golden, CO 80401

Alan S. Hirshberg
Booz-Allen Applied Research
4330 East-West Highway
Bethesda, MD 20014

Arjun Makhijani
Foundation for Research in Community Health
Dhokawade, P.O. Awas
Taluk Alibag, District Kolaba
Maharashtra, India

Alan D. Poole
Institute for Energy Analysis
P.O. Box 117
Oak Ridge, TN 37830

Marc H. Ross
The Harrison M. Randall Laboratory of Physics
University of Michigan
Ann Arbor, MI 48104

William A. Shurcliff
19 Appleton Street
Cambridge, MA 02138

# Contributors

Bent Sørenson
Niels Bohr Institute
University of Copenhagen
Blegdamsvej 17
DK-2100 Copenhagen Ø
Denmark

Frank Von Hippel
School of Engineering/Applied Science
Center for Environmental Studies
Princeton University
Princeton, NJ 08540

Robert H. Williams
School of Engineering/Applied Science
Center for Environmental Studies
Princeton University
Princeton, NJ 08540

Martin Wolf
Department of Electrical Engineering
University of Pennsylvania
Philadelphia, PA 19174

Clarence Zener
Carnegie-Mellon University
Schenley Park
Pittsburgh, PA 15213

Absorbing sheets, 22–23
Accidents
  nuclear, 90
  windmill, 138
Active-type solar heating, 17–31
  cost of, 25–26
  durability of, 24–25
  efficiency of, 18–24
  principles of operation of, 17–18
  retrofit problems in, 26–27
  standards for, 27–28
Agrément systems, 54
Agriculture
  biomass production and, 155, 156,
    158, 225, 227
  draft energy in, 185–186
  photosynthetic production in,
    149–150
  Third World energy needs and,
    184–191
Air conditioning, 205, 207–208, 231
Air-type collectors, 25
Aleutians, 7
Aluminum
  efficiency in production of, 206
  solar collectors with, 23, 24, 25
American Physical Society, 203
Ammonia, 98, 105
Anderson, James H., 96
Anderson, J. Hilbert, 96
Antal, Michael J., Jr., 163, 169–178
Antifreeze, 25
Ash content, 145
Atmospheric pressure system, 125–126
Automobiles, 207

Barstow, California, 75
Beck, E. J., 98
Beckman, W. A., 20
Bihar, India, 185
Biogasification
  animal manure for, 162
  costs and, 163, 174, 225
  plant biomass for fuel in, 150–151,
    152–153, 161
  solar furnaces for, 172
  Third World energy needs and,
    187–191
  water hyacinth and, 199
Biomass conversion processes
  bog harvesting in, 198–199

electric power plant fuel from,
  153–154
hydrogen in, 171
management in, 156–158
methane from fermentation in,
  152–153
ocean farming in, 158–160
organic materials in, 160–162, 234,
  235
production scale in, 154–160
steam reforming process in, 171–172
terrestial system for, 155–158
Bliss, Raymond W., 26, 33–49
Bogs, 198–199
Boilers, 100
Boston, 36–40
Brayton engines, 80
Breeder reactor programs, 88, 223
Building industry
  adoption of solar heating and, 52
  building codes and, 54
  economic competitiveness in, 56–61
  low-interest loans in, 59–60
  organizational structure of, 55–56
  risk-reducing policies in, 61–63
  tax incentives in, 59

Cadmium sulfide, 109, 110, 115, 120
California, 67
California Coastal Zone Conservation
  Commission, 58
Capital costs
  of biogasification, 174–175
  nuclear power plants and, 101–103
  of ocean-based solar plants, 5
  photoelectric cells and, 3
  photovoltaic energy conversion and,
    112
  residential use of solar energy and, 3–4
  solar energy and, 224–225
  wind energy and, 10, 132, 134
Caputo, Richard S., 73–93
Carbon, 187
Carbon dioxide, 13, 90
  biomass production and, 160, 171–172
  fossil fuels and, 222–223
  methane conversion and, 152
  ocean-based solar plants and, 7
  photosynthesis and, 11, 146, 147
  pyrolysis and, 171
Carbon monoxide, 171

# Index

Caribbean area, 100, 105
Carnegie-Mellon University, 98
Central receiver power plants, 76-79, 81
Char, 171
Charoil, 162, 234
Chemical plants, 232
Chickens, 199
Chlorine, 10-11
China, 195
Claude, Georges, 96
Clearing houses, 60-61
Climate
  fossil fuels and, 90, 222-223
  ocean-based solar plants and, 5-6, 95
  wind energy and, 139
Clothing, 193-195
Coal gasification
  health effects of, 90
  pricing policies with, 57
  solar plants compared with, 83
Coal power plants, 84
  climate and, 222-223
  health effects of, 88-90
  research and development costs for, 88
  solar plants compared with, 83
Coconut palm, 199
Codes, building, 52, 54, 56
Collection areas
  community, 226
  solar energy and, 224
Collectors
  air-type, 25
  durability of, 24-25
  efficiency of, 21-25
  hot water heating in, 30
  power towers and, 172-173
  pricing policies and, 63-64
  solar power plants and, 76
  standard condition of collector coolant
    and, 21
  standard condition of irradiation in, 21
Colorado River, 74
Commercial buildings
  savings in air conditioning in, 208
  solar units in, 30
Commercialization of technologies, 197
Compressed air, 131, 133
Congress, 59, 60
Conservation programs, 203, 213-216
Construction (housing)
  barriers to adoption of solar heating in,
    52
  Boston study of efficiency in, 36-42

building codes and, 54
costs and, 225
direct solar heating and, 42-49
financial constraints in, 54-55
improvements in, 33-49
incentive programs in, 59, 64-69
low-interest loans for, 59-60
risk-reducing policies for, 61-63
savings from, 47-48
Construction (power plants), 79, 88
Converters, 109-110
Cooking
  fuel for, 196
  solar cookers for, 199
  stoves and, 199
  Third World needs for, 181, 191-193
Coolant in collectors, 21
Cooling techniques, 74-75
Copper, 23, 24, 25
Copper sulfide, 109, 115, 120
Coppicing, 196
Costs
  active-type solar heating, 25-26
  assessing impact of policies on, 63-64
  biomass as fuel and, 150, 153, 225
  coal technology and, 88
  competitiveness of energy sources and,
    56-61
  conservation programs and, 203
  construction of houses and, 35, 36-42,
    47
  conventional power plants comparison
    for, 81-84
  cooking stoves and, 193
  cooling in solar power plants and,
    74-75
  hot water heating and, 30
  improvements in efficiency and, 24
  incentive programs and, 66
  methane conversion and, 150-151,
    152-153, 189
  nuclear energy compared with wind
    energy for, 136
  nuclear-OTEC system comparison for,
    101-102
  as obstacle to acceptance of solar heat-
    ing, 51
  optical concentrators in photovoltaic
    systems and, 112, 113
  OTEC power system and, 95-96,
    99-100
  payback period of, 51
  photosynthetic energy and, 162-163

photovoltaic energy system and,
111–112, 115–116
solar power plants and, 76–81, 221
solar water heating and, 195
transmission of solar electricity and,
234
village public utilities and, 196–197
wind energy and, 131, 132, 134
windmills and, 186
*see also* Capital costs
Craft unions, 56
Credits, tax, 59–60, 64
Crops. *See* Agriculture
Cuniff, Charles, 36
Czochralski method, 119

Daniels, Farrington, 172
Danube delta, 156, 158
D'Arsonval, Jacques, 96
Davis, E. S., 58, 63
Decentralization of electricity genera-
tion, 225
Demonstration projects, 26, 62
Department of Energy (DOE), 75, 95,
103
Department of Housing and Urban
Development (HUD), 27, 60
Depreciation, 59
Deregulation of gas, 56–57, 58
Desalination, 1
Design
capital costs and, 3
government grants for, 28–29
patents and secrecy of, 28
regional differences and, 55
solar heating in housing and, 4, 43–44
Developing nations, 165
Development strategies, 179–239
Diesel generator set, 234
Direct solar heating, 42–49
*Direct Use of the Sun's Energy,* (Daniels),
172
Distributed receiver power plants, 79–80
Dow Chemical Company, 208
Draft energy, 185–186
Dry towers, 74, 83
Duffie, J. A., 20
Durability of collectors, 24–25

Economic conditions
adoption of solar energy and, 95–96
building for tomorrow and, 29

competitiveness in building industry
and, 56–61
land-based solar plants and, 5
ocean-based solar plants and, 7
photovoltaic energy conversion and,
114–115
power production and, 101–104
power tower pyrolysis and, 174–176
retrofit problems and, 27
savings through construction and,
47–48
wave power and, 10
zero energy growth and, 215–218
*see also* Capital costs; Costs; Scale
economies
Edge-defined film growth (EFG), 119
Efficiency
active-type solar heating and, 18–24
as clear-cut concept, 20–22
conservation for long term and, 213
cooking stove, 193
desirability of increasing, 23–24
energy resources and, 203–219
first-law, 205
housing construction and, 36–40
improvements in existing plants and,
204–206
increases in, 22–23
nuclear versus wind energy and, 136–
137
OTEC power system and, 99
period of time and, 21
photosynthetic energy and, 11, 145,
147–148
photovoltaic energy conversion and,
111
power tower fuel production and, 173
second-law, 205–206, 214
standard condition of irridiation and,
21
water hyacinth conversion and, 235
Electric power plants
biomass as fuel for, 153–154
collector system and dependency in,
25
cost in, 134
decentralization of, 225
energy use projections and, 214–215
household consumption and, 3
housing construction and, 39–40
land-based solar, 4–5
ocean-based solar, 5–7

Electric power plants (continued)
 pricing policies for, 67, 232–233
 solar, 73–141, 233–234
 wind energy in production for,
  130–131
Electrolysis, 132, 234–235
Energy Conservation and Conversion Act
  of 1975, 59
Energy Extension Service, 61–62
Energy Policy Project, 203, 216
Energy requirements, 229–230
Energy Research and Development
  Administration (ERDA), 1, 24, 28,
  57, 60, 61, 103, 121, 233
Engine design, 207
Engineering, 10
England, 156
Environmental factors
 biomass production and, 156
 coal and, 222–223
 conservation programs and, 203
 ocean-based solar plants and, 7
 power plant comparisons for, 90
 solar satellites and, 10–11
 Third World energy needs and, 198
 wind energy and, 138–139
Ethiopia, 191
Ethylene glycol, 25
Eutectic salt, 27
Evaporation, 1

Federal Energy Administration (FEA),
  57, 60
Federal Housing Administration (FHA),
  36–45
Feedstock, 152, 159
Fermentation, 152–153
Fertilizers
 biomass production and, 159, 225
 Third World energy needs and, 184,
  185
Fetkovich, J., 98
Financing
 building industry and, 54–55, 56
 low-interest loans in, 59–60
 tax incentives in, 59
Fish, 199
Florida, 67, 231
Fluorocarbon plastics, 22, 23
Flywheels, 131, 138
Food production, 12
Ford Foundation, 203, 216

Forest resources
 energy production with, 155, 156,
  162, 163
 village wood supplies and, 192
Fossil fuels
 climate and, 90, 222–223
 conservation programs and, 203
 heating with, 2
 solar plants compared with, 83
 see also Coal power plants
Fuel cells, 133
Furnaces
 household, 204, 205
 solar, 163–165
Fusion, nuclear, 221, 223

Gallium arsenide, 109, 110, 111,
  115–116
Gandhi, Mohandas, 194
Gangetic plain, 191, 196
Gas. See Natural gas
Gasification. See Biogasification;
  Coal gasification
Gedser, Denmark, 134, 136
Genetic engineering, 198
Glass, in collectors, 22
Global Marine Development, Inc.,
  100
Government
 grants from, 28–29
 incentive packages from, 64, 215
 photovoltaic energy system and, 111,
  120–121
 standards from, 27–28
 village public utilities and, 195
Grants, 28–29
Great Plains, 227
Greenhouse effect, 223
Gulf of Mexico, 6, 105, 227
Gulf Stream, 6–7, 98, 100

Hawaii, 7
Hay, H. R., 26
Health effects of solar energy, 88–90
Heat exchange systems, 6, 100
Heating systems
 active-type, for housing, 17–31
 barriers to use of, 51–69
 Boston housing construction study of,
  36–42
 building codes and, 54
 clothing and, 193–195

in commercial buildings, 30
energy balance equations for, 33-34
financing constraints and, 54-55
furnaces in, 204, 205
hydrogen production for, 133
passive-type, 17
photovoltaic energy conversion for, 108
potential savings in, 207
retrofit problems in, 26-27
secondary-law efficiency for, 206
solar, 15-69, 231
standards for, 27-28
Third World needs for, 181, 191-193
water (see Water heating)
wind energy in, 131-132
Heat loss, 35
Heat pumps, 30
household furnaces and, 204
improved technology in industries for, 216
savings in space heating and, 207
wind energy and, 132
Heronemus, William, 98
Hirshberg, Alan S., 51-69
Hottel, H. C., 26
Hot water heating. See Water heating
Housing
active-type solar heating in, 17-31
air infiltration in, 34, 35
building codes and, 54
capital costs of solar energy and, 224-225
conservation programs and, 203
direct solar heating in, 42-49
efficiency study of, 36-40
energy balance equation in, 33-34
financial constraints in solar heating in, 54-55
fuel-saving principles in, 35-42
improved building techniques for, 33-49
incentive programs in, 64-69
retrofit problem in, 26-27
risk-reducing policies in, 61-63
HUD (Department of Housing and Urban Development), 27, 60
Hyderabad Engineering Research Laboratories, 193
Hydrogen, 7
electrolysis of water and, 234
methane production and, 169-171

power tower fuel production and, 173
solar furnaces in biomass conversion and, 163-165
steam reforming process and, 171-172
wind energy for, 132-133
Hydrogen sulfide, 187
Hydropower pilot projects, 199

Incentive programs, 64-69, 215
Income tax, 59
India, 185, 192, 193, 196, 198
Indian Institute of Science, 187, 198
Industrial parks, 154
Industries
energy conservation technology and, 216
potential savings for, 208
solar-heating, 17
Infrasound waves, 139
Institute for Energy Analysis, 104, 158
Insulation in housing, 3
direct solar heating and, 42-47
housing construction studies and, 36-38
savings from, 215
Interest rates, 54
Intrastate natural gas, 56
Investment tax credit, 59
Iraq, 191
Irrigation
biomass production and, 155
solar water heating and, 195
Third World needs and, 184
wind energy for, 133

Jet Propulsion Laboratory, 62, 79
Johns Hopkins University, 105

Labor unions, 186
Land use
biomass production and, 153-154, 155-158, 161, 235
coal plants and, 88, 222
photovoltaic energy conversion and, 114-115
solar power plants and, 5-6, 7-8, 73-93
solar water heater and, 195
woodlots in villages and, 196
Legislation, 59
Legumes, 185
Lending institutions, 55

Lighting, 207-208
Light water reactors, 83, 84, 88
Lignin, 152
Loans, 29, 59-60, 64, 215
Lockheed Company, 98, 103-104
Los Alamos Scientific Laboratory, 173, 174

Maintenance costs, 136
Majarashtra, India, 193
Makhijani, Arjun, 181-201
Management, in biomass production, 156-158
Manure
  energy conversion with, 162, 163, 189
  water hyacinth cycle and, 199
Market penetrating model, 62-63
Mechanization in agriculture, 186
Methane, 12
  biomass fermentation to, 152-153, 169, 189
  conversion costs for, 150-151, 189
  kelp farming in production of, 159-160
  sewer gas production of, 199
  Third World energy needs and, 187-191
Methanol, 151, 169, 171
Mexico, 185
Middle Atlantic states, 74
Miniaturization, 117
Mining, 88
Morgan, A. E., 26

National Academy of Engineering, 103-104
National Aeronautics and Space Administration (NASA), 235
National Bureau of Standards, 27
National Construction Indicator, 64
National Science Foundation (NSF), 24, 28, 57, 76, 96-98, 103
Natural gas, 226-227
  competitiveness of, 56
  construction of power plants using, 101
  deregulation of, 56-57, 58
  pricing of, 67
Nitrogen, 185, 187, 191
Noise, in wind energy, 138-139
Nova Scotia, 149

Nuclear power plants
  environmental factors in, 198
  health effects of, 88-90
  OTEC power system compared with, 101-103
  reliability of, 101-103
  social resistance to, 84-85
  solar plants compared with, 84-85
  wind energy compared with, 134-138
Nutrient supply, 149, 158

Oceans
  biomass production and, 158-160
  OTEC power system and, 95-105
  solar plants and, 6-8, 227-229
  wave power in, 10
Oil-fired power plants, 134, 137
Oil heating, 39-40
Oil shale, 169
Olds, F. C., 101
Optical concentrators, 112-113
Organic materials. See Biomass conversion process
OTEC power system, 95-105
  future role of, 104-105
  objections to, 98-100
  operation of, 95-97
Oxidation, through wind energy, 133
Oxygen, 146
Ozone layer, 10-11, 13

Paper mills, 232
Parabolic dish collectors, 76, 79-80
Particulate emissions, 90
Passive-type solar heating, 17, 29
  design requirements for, 43-44
  housing construction and, 42-49
Patents, 28
Payback period, 51
Peak load pricing, 231-232
Performance, and efficiency, 205
Peru current, 149
Phosphorus, 158, 159-160
Phosphorus pentoxide, 191
Photo-electric cells, 2, 3, 10
Photosynthetic energy, 145-168
  biomass fermentation to methane in, 152-153
  bog harvesting and, 198-199
  fundamentals of, 146-150
  limiting factor in, 145

plant biomass as fuel in, 150–154, 234, 235
potential impact of, 162–165
production scale of biomass in, 154–160
as solar energy conversion process, 1, 11–12
Third World energy needs and, 181, 182–184, 191
Photovoltaic solar energy conversion, 107–123
commercial feasibility of, 120–122
conditions for, 107–108
converters in, 109–110
cost of, 111–112
new industry connected with, 113–120
optical concentrators in, 112–113
solar array production in, 116–120
Pigs, 199
Plastic collectors, 22
Plutonium, 223
Pollution
coal and, 222
health effects of, 88–90
ocean-based solar plants and, 7
photosynthetic energy and, 145
solar energy systems and, 13
solar satellites and, 10–11
Poole, Alan D., 12, 145–168, 169, 172, 174, 177
Potassium oxide, 191
Power dish collectors, 76
Power plants
conventional, 81–84
cooling techniques in, 74–75
costs of, 76–81
health effects of, 88–90
regional constraints on, 73–74
social costs of, 85–92
Power tower system, 172–174
Precipitation, 148, 155
Pricing policies, 67
energy sources and, 56–58
peak load, for electricity, 231–232
photovoltaic energy system and, 112
Princeton University, 173
Production
mass production and costs in, 26
optical concentrators and, 112–113
OTEC power system and, 100
patents and secrecy in, 28

photovoltaic energy conversion and, 108, 111, 112–113, 116–120
problems in future for, 29–30
standards and, 28
Third World energy needs and, 197
Project BASE, 62
Project SAGE, 57
Property tax, 59
Propulsion, with wind energy, 133
Protein feed, 152
Public Law 93–409, 60
Puerto Rico, 7
Pumps
heat (*see* Heat pumps)
water, 133, 186
Pyrolysis, 162
plant biomass conversion with, 151–152
power tower and, 172–174

Rainfall, 1, 148, 155
Rankine steam cycle, 75
Recreation, 156
Reddy, A. K. N., 198
Reed culture, 158
Reforestation, 192
Regional systems, 226–229
Reradiation energy, 23
Research and development
coal technology and, 88
photovoltaic energy conversion and, 121
Third World energy needs and, 197–200
Retrofit problems, 26–27
Ridgeway, S., 98
Rio Grande, 74
Ross, Marc H., 203
Rotation of crops, 196
Rotation of trees, 158
Rumania, 156

Sales tax, 59
Sargasso Sea, 149
Satellites, 10–11, 107
Saunders, N. B., 26
Savings
conservation programs and, 207–209
housing construction and, 47–48
payback period and, 51
Scale economies
biomass production and, 153, 158

Scale economies (continued)
distributed receiver power plants and, 79
manufacturing and, 63
organic wastes in biomass production and, 160-161
photovoltaic system and, 108
solar energy plants and, 225-226
Third World energy needs and, 197-198
Schoen, Richard, 54
Scott, J. A., 67
Seas. *See* Oceans
Second-law efficiency, 205-206, 214
Secrecy, 28
Semiconductor industry, 116
Sewage gas, 199
Sewage treatment ponds, 235
Ship propulsion, 133
Shurcliff, William A., 17-31
Silicon, 108-111, 115-120
Simplicity in collectors, 25
Social factors
power plant comparisons and, 90
resistance to nuclear power plants and, 84-85
solar power plants and, 85-92
Sodium sulfate dekahydrate, 27
Sodium thiosulfate pentahydrate, 27
Solar array production, 116-120
Solar cells, 107, 110-111
Solar cookers, 199
Solar Energy Industries Association (SEIA), 60
Solar furnaces, 163-165
Solar Heating and Cooling Demonstration Act of 1975, 60, 62
Solid state system, 107, 108
Sørensen, Bent, 125-141
Southeastern states, 7, 154, 155
Southern California Edison Company, 62
Southern California Gas Company, 57
Southwestern states, 7, 73, 92, 227
Space heating. *See* Heating systems
Space probes, 107
Space program, 108
Space shuttles, 10-11
Standards
efficiency and, 21
heating systems and, 27-28

States
grants from, 29
incentive package, 64
Steam for industrial use
biomass as fuel for, 153-154, 171-172
land-based solar plants for, 5
potential savings with, 208
Stirling engines, 80
Storage systems
active-type solar heating with, 17-18, 21-22, 24
capital costs in residential usage and, 3
direct solar heating and, 43-44
distributed receiver power plants with, 79
eutectic salt in, 27
heating systems and, 231
methane and, 189
ocean-based solar plants and, 6
OTEC power system and, 95
photovoltaic energy conversion and, 108, 111
solar power plants and, 73, 224
wind energy and, 9-10, 130, 131, 133, 138
Sulfur, 145
Sulfur oxides, 88
Synthetic fuels, 169

Tar sands, 169
Tax abatements, 59, 63
Tax incentives, 29, 52, 59-60, 63, 64, 215
Telkes, M., 26, 27
Thermo Electron Corporation, 208-209
Third World countries
agricultural energy needs of, 184-191
clothing use in, 193-195
energy use in, 182
physical needs of, 181-182
research into needs of, 197-200
solar strategies and rural development in, 181-201
village public utilities and, 195-197
Thomas, Percy, 9-10
Thomason, Harry E., 4, 21, 26
Thorton, Ray, 61
Trace minerals, 158
Tracking systems
community collection systems and, 226

efficiency and, 22, 23
optical concentrators in photovoltaic
   systems and, 113
Transmission of energy
   ocean-based solar plants and, 5-7
   plant biomass as fuel and, 150, 154
   solar power plants and, 75, 227, 234
Transmissions, automobile, 207
Transportation systems, 151
Trichlorosilane, 116
Troombe, F., 26
TRW Company, 98, 100, 103-104

Unions, 56, 186
Uranium, 223
Utilities' ownership, 57-58, 195-197

Ventilation systems, 207-208
Villages
   fuel wood supply in, 192
   public utilities in, 195-197
Von Hippel, Frank, 1-14, 95-96,
   221-239

Washing clothes, 194-195
Washington, D.C., 3
Waste disposal, nuclear, 90
Waste heat
   biomass conversion systems and, 154
   land-based solar plants and, 5
   potential savings from, 208
Waste resources
   biomass conversion from, 160-162,
      163, 169, 171, 225
   water hyacinth in, 199
Water
   hydrogen production and, 234-235
   photosynthesis and, 146, 148, 155
   solar plant cooling with, 5, 74
   Third World need for, 186
Water heating, 30, 231
   competitiveness in, 56
   photovoltaic energy conversion for,
      108
   Third World needs for, 184
   washing clothes and, 194-195
Water hyacinths, 199, 235
Wave power, 10
Weapons materials, nuclear, 88, 90
Weather, 22, 223
Weather Bureau, 36

Westinghouse Company, 98, 104
Wetlands, 156
Wilderness areas, 156
Williams, Robert H., 1-14, 95-96,
   145-168, 169, 172, 174, 177, 203,
   221-239
Wind energy, 7-9, 125-141
   applications of, 130-134
   areas of world favorable to, 126-130
   atmospheric pressure system and,
      125-126
   configuration of windmills and,
      126-128
   electricity production and, 130-131
   environmental impact of, 138-139
   heat production with, 131-132
   hydrogen production with, 132-133
   nuclear energy compared with,
      134-137
   number of windmills needed for,
      8-9
   Third World use of, 184, 186-187
   transmission problems in, 7
Windmills, 125, 227
   accidents involving, 138
   configuration of, 126-128
   heat production with, 131
   infrasound waves from, 139
   noise from, 138-139
   number of, 8-9
   power fluctuations with, 130-131
   Third World agricultural needs and,
      186-187
Window area
   direct solar heating and, 4, 42-47
   housing construction studies and, 35,
      36, 39, 40, 41-42
Wolf, Martin, 107-123
Wood, 181, 191, 192, 193, 196

Xenophon, 35

Zener, Clarence, 95-105
Zero energy growth, 209-213, 216-218

## DATE DUE

| | | | |
|---|---|---|---|
| AP 29 '80 | | | |
| OC 8 '80 | | | |
| MR 2 3 '81 | | | |
| AP 2 2 81 | | | |
| MR 3 0 '82 | | | |
| | | | |
| AP 2 8 '82 | | | |
| DE 6 '82 | | | |
| AP 24 '83 | | | |
| NOV 1 7 '86 | | | |
| DEC 7 87 | | | |
| | | | |
| | | | |
| | | | |
| | | | |
| | | | |
| | | | |
| GAYLORD | | | PRINTED IN U.S.A |